Introduction
to a
Submolecular
Biology

Introduction
to a
Submolecular
Biology

ALBERT SZENT-GYÖRGYI

The Institute for Muscle Research
at the Marine Biological Laboratory
Woods Hole, Massachusetts

1960

ACADEMIC PRESS
New York and London

ACADEMIC PRESS INC.
111 FIFTH AVENUE
NEW YORK 3, N. Y.

United Kingdom Edition
Published by
ACADEMIC PRESS INC. (LONDON) LTD.
17 OLD QUEEN STREET, LONDON S.W. 1

Library of Congress Catalog Card Number 60-14265

PRINTED IN THE UNITED STATES OF AMERICA

Acknowledgments

WHEN SETTLING DOWN TO WRITE THIS BOOK my thoughts turn gratefully to my associates who have given me part of their lives, sharing my work, hopes and disappointments.

My thanks go to those who gave me their confidence and help, making this work possible. In the first place, I am thinking of the Commonwealth Fund which helped me from the beginning in spite of my warning that it was supporting a gamble. I think of the Muscular Dystrophy Associations, the American Heart Association, the Association for the Aid of Crippled Children, and the Cerebral Palsy Association. My thoughts also wander back to those who helped me earlier, in my more troubled days, notably Armour and Company.

Last but not least, I am thinking of the people of the USA who gave me a new home, accepting me as one of their own, supporting my work through the National Institutes of Health, Bethesda, Maryland, and the National Science Foundation, Washington, D. C.

According to the A.L.A. Interlibrary Loan Code

Date of request:

Call No.

For use of   Robert Saunders   Status   Faculty   Dept.

Author (or periodical title, vol. and year)

     Szent-Gyorbyi, Albert, 1893–

Title (with author & pages for periodical articles) (Incl. edition, place & date)   ☐ This edition only

     Introduction to submolecular biology. N.Y. Academic

     Press, 1960

Verified in (or source of reference)   BIC

If non-circulating please supply ☐ Microfilm ☐ Hard copy if cost does not exceed $ _____

     Interlibrary Loans
     Philadelphia College of Pharmacy
     & Science

Note: The receiving library assumes responsibility for notification of non-receipt

AUTHORIZED BY: _____
(FULL NAME)   Title _____

D

NOTICE OF RETURN
RETURN SEPARATELY

REPORTS: Checked by _____

SENT BY: ☐   Library rate ☐

Charges $ _____ Insured for $ _____

Date sent   10/23/74

DUE   10/24/74

RESTRICTIONS: ☐ For use in library only

☐ Copying not permitted ☐ _____

NOT SENT BECAUSE:

☐ Non circulating   ☐ In use
                      ☐ Not Owned

Estimated Cost of: Microfilm _____
                  Hard copy _____

BORROWING LIBRARY RECORD:

Date received   10/25/74

Date returned   10/6/74

By ☐   Library rate ☐

Postage
enclosed $ _____ Insured for $ _____

RENEWALS:

Requested on _____

Renewed to _____
    (or period of renewal)

Abbreviations

~	High-energy bond (mostly phosphate bond).
A	Acceptor
ADP	Adenosine diphosphate
AMP	Adenosine monophosphate
ATP	Adenosine triphosphate
D	Donor
DPN	Diphosphopyridine nucleotide
EA	Electron affinity
EV	Electron-volt
ESR	Electron spin resonance
FAD	Flavine-adenine-dinucleotide
FMN	Flavine-mono-nucleotide
IP	Ionization potential
IR	Infrared
P	Phosphate
TPN	Triphosphopyridinenucleotide
UV	Ultraviolet

Contents

Part One

General
Considerations

I

Introduction

THE BASIC TEXTURE OF RESEARCH CONSISTS OF dreams into which the threads of reasoning, measurement, and calculation are woven.

This booklet is the reincarnation of my "Bioenergetics"[1] which was hardly more than a dream. So, it was a surprise to me to find it translated into Russian by the Academia Nauk USSR, and to find that the introduction was written by A. Terenin, a leading figure of Soviet science. Two years later, in the fall of 1959, the Atomic Energy Committee organized a meeting, in Brookhaven, on "Bioenergetics," which gave to the problem the status of a more or less well-defined field of inquiry.

Since 1957 my thoughts have assumed a somewhat more definite form and even if I am unable to present final solutions, I am capable, at least, of asking a few questions more intelligently, weaving in a few threads of measurement, reasoning, and calculation. All the same, it is not without a great deal of anxiety that I

---

[1] "Bioenergetics," Academic Press, New York, 1957.

1

publish this booklet which will be the last instance of the repetitive pattern of my being driven into fields in which I was a stranger. I started my research in histology. Unsatisfied by the information cellular morphology could give me about life, I turned to physiology. Finding physiology too complex I took up pharmacology, in which one of the partners, the drug, is of simple nature. Still finding the situation too complicated I turned to bacteriology. Finding bacteria too complex I descended to the molecular level, studying chemistry and physical chemistry. Armed with this experience I undertook the study of muscle. After twenty years' work, I was led to conclude that to understand muscle we have to descend to the electronic level, the rules of which are governed by wave mechanics. So here again, I was driven into a dimension of which I had no knowledge. In earlier phases I always hoped, when embarking on a new line, to master my subject. This is not the case with quantum mechanics. Hence my anxiety.

I have not referred to my personal history as if, in itself, it would be of any importance. I have referred to it because a most important question hinges on it: should biologists allow themselves to be steered away from this electronic dimension because of their being unfamiliar with the intricacies of quantum mechanics? At present, the number of those who master both sciences, biology and quantum mechanics, is very small. Maybe it will never be very great owing to the limited nature of human life and brain. Both sciences claim a whole mind and lifetime. So, at least for the present, developments depend on some sort of hybridization.

In my opinion, at least temporarily, the best solution does not lie in the biologists crossing over into physics, and *vice versa*, but in the collaboration of biologist and physicist. For this it is not necessary for the biologist to acquaint himself with the intricacies of wave mechanics. It is sufficient to develop a common language with the physicist, get an intuitive grasp of the basic ideas and limitations of quantum mechanics, to be able to isolate problems for the physicist and understand the meaning of his answer. Similarly, the physicist had better stay on his side of the fence rather than become, perhaps, a second-rate biologist. If, for example, as a biologist, I am interested in energy levels of a substance, and am told that the highest orbital of a substance has a k value of, say, 0.5, I can start from this point. It is sufficient for me to know what $k = 0.5$ means, and there is no need for me to know exactly how the value was arrived at. In exchange, I can bring substances to the notice of the physicist, the k value of which might be of special importance.

There is only one warning I would like to give to the biologists who venture into physical problems. There is a basic difference between physics and biology. Physics is the science of probabilities. If a process goes 999 times one way, and only once another way, the physicist will not hesitate to call the first *the* way. Biology is the science of the improbable and I think it is on principle that the body works only with reactions which are statistically improbable. If metabolism were built of a series of probable and thermodynamically spontaneous reactions, then we would burn up and the machine would run down as a watch does if deprived of its regulators. The reactions are kept in hand by being statistically improbable and made possible by specific tricks which may then be used for regulation. So, for the living organism, reactions are possible which may seem impossible, or, at least, improbable to the physicist. When Tutankhamen's grave was opened, his breakfast was found unoxidized after three thousand years. This represents the physical probability. Had His Majesty risen and consumed his meal this would have been burned in no time. This is biological probability. His Majesty, himself, must have been a very complex and highly ordered structure of nuclei and electrons with a statistical probability of next to zero. I do not mean to say that biological reactions do not obey physics. In the last instance it is physics which has to explain them, only over a detour which may seem entirely improbable on first sight. If Nature wants to do something, she will find a way to do it if there is no contradiction to basic rules of Nature. She has time to do so.*

All this makes the relationship of physicist and biologist rather touchy. The biologist depends on the judgment of the physicist, but must be rather cautious when told that this or that is improbable. Had I always accepted the physicist's verdict as the last word, I would have given up this line of research. I am glad I did not. One can know a good theory from a bad one by the former's leading to new vistas and exciting experiments, while the latter mostly gives birth only to new theories made in order to save their parents.

* Living Nature also often works with more complex systems than the physicist uses for testing his theories.

4

Since I am working on my present line everything seems more colorful and I am even more eager to get to my laboratory in the morning than ever before.

The biologist, embarking on this line, is neither without help nor without encouraging examples. C. A. Coulson, in the first volume of the "Advances in Cancer Research" (Academic Press, New York, 1953) has written an admirably clear unmathematical review of the basic concepts of quantum mechanics. In the third volume of the same reviews (1955) A. and B. Pullman wrote equally clearly about complex indices. Those who like to read French may find a more comprehensive review by the same authors in "Cancérisation par les substances chimiques et structure moléculaire" (Masson, Paris, 1955). For details and some mathematics one may apply to B. and A. Pullman's "Les théories électroniques de la chimie organique" (Masson, Paris, 1952). The other extreme, a very brief and popular summary, is found in B. Pullman's "La structure moléculaire" (Presses Universitaires France, 1957). Naturally, L. Pauling's "The Nature of the Chemical Bonds" (Cornell University Press, 1948) should not be missing from one's desk, nor Th. Förster's "Fluoreszenz Organischer Verbindungen" (Göttingen 1951). Its next (English) edition is eagerly awaited.*

As encouraging examples of the fruitfulness of this field I would like to mention the bold pioneering stud-

---

* In spite of these splendid contributions it is difficult to deny that an up-to-date comprehensive treatment, written especially for the biologist, in a possibly unmathematical language, is badly needed.

ies of A. and B. Pullman which started with the study of the electronic structure of carcinogenic hydrocarbons, in which correlations of electronic structure and carcinogenic power were established, which may be one of the first major steps toward understanding cancer. If this booklet does not contain a chapter on this subject this is because the Pullmans have, themselves, given an account of their work which could not be equaled in clarity. The same authors have broken ground also in various other fields of quantum mechanical biology, establishing electronic indices for many biologically important catalysts,[2-4] and venturing even into the field of enzymes[5] and high-energy phosphate bonds.[6]

While the quoted examples may encourage physicists interested in biology, B. Commoner[7-10] and his associates may be quoted for their pioneering studies in electron spin resonance to cheer biologists interested in submolecular phenomena. It was this work

[2] B. Pullman and A. Pullman, *Proc. Natl. Acad. Sci. U.S.* **44**, 1197, 1958.

[3] B. Pullman and A. Pullman, *Proc. Natl. Acad. Sci. U.S.* **45**, 136, 1959.

[4] B. Pullman and A. M. Perault, *Proc. Natl. Acad. Sci. U.S.* **45**, 1476, 1959.

[5] A. Pullman and B. Pullman, *Proc. Natl. Acad. Sci. U.S.* **45**, 1572, 1959.

[6] B. Pullman, *Radiation Research*, 1960.

[7] B. Commoner, J. Townsend, and G. E. Pake, *Nature* **174**, 689, 1954.

[8] B. Commoner, J. J. Heise, B. B. Lippincott, R. E. Norberg, J. V. Passoneau, and J. Townsend, *Science* **126**, 3263, 1957.

[9] B. Commoner, B. B. Lippincott, and Janet V. Passoneau, *Proc. Natl. Acad. Sci. U.S.* **44**, 1099, 1958.

[10] B. Commoner and B. B. Lippincott, *Proc. Natl. Acad. Sci. U.S.* **44**, 1110, 1958.

which led to the first direct experimental evidence for the participation of free radicals in the one-electron enzymic electron transfer, as outlined in the classic studies of L. Michaelis.[11]

[11] L. Michaelis, Fundamentals of Oxidation and Reduction, *In* "Currents in Biochemical Research." D. E. Green (Ed.), Interscience Publ., New York, 1946.

II

Why Submolecular Biology?
The Problem is Stated

LOOKED AT FROM A DISTANCE, THE HISTORY OF
biochemistry seems to be but a series of astounding
successes, a blaze of glory. The rate of progress shows
no decrement and it looks as if soon we could strike
out "don't know" from our vocabulary, altogether.
Why, then, talk about "submolecular biology" until
molecular biochemistry has run its full course?

There is no doubt about these successes. All the
same, if one does not allow oneself to be blinded by
them and approaches biochemistry with a dark-
adapted eye, big gaps in our knowledge become evident. Let us consider some of the main problems of
chemical biology, starting with metabolism. Biochemistry has unraveled the complex cycles of intermediary
metabolism and has shown that the main object of this
metabolism is to prepare the foodstuffs for their final
oxidation in which their energy is used to couple one
molecule of phosphate to ADP, producing, thereby,
ATP (Fig. 19). In this process the energy of the food-
stuff is translated into the energy of the terminal
"high-energy phosphate bond," $\sim$, of a very specific

9

molecule. It is only in this form that the energy of the foodstuffs can serve as fuel for the living machinery and drive it. This "oxidative phosphorylation" is thus the central event of metabolism. Its mechanism is completely unknown.

We are equally ignorant about the reversal of this process, the release of the energy of the $\sim$ of ATP. How these $\sim$'s drive life, how their energy is translated into various forms of work, w, be they mechanical, electrical, or osmotic, we do not know, although this transformation may be the most central problem of biology. We know life only by its symptoms and what we call "life" is, to a great extent, but the orderly interplay of these various w's; since the dawn of mankind, death has been diagnosed, mostly, by the absence of one of these w's, expressing itself in motion. We do not know how motion is generated, how chemical energy is transformed into mechanical work.

Physiology has shown that the various functions of our body are regulated and coordinated by hormones and the biochemist will proudly show the row of vials containing these mysterious hormones mostly in the form of nice, crystalline powders, some of which might have been prepared synthetically. The same is true for the various vitamins, the catalogue of which seems near completion. The biochemist will be able to give us the structural formula of most of these substances. The really intriguing problem, however, is not what these substances *are*, but what they *do*, how they act on the molecular level, how they produce their actions. There is no answer to this question. The same holds true also for the majority of drugs.

As to the living machinery itself, the biochemist will

10

tell you that its central parts are proteins, nucleic acids, and nucleoproteins. He will point out the great progress made in the structural analysis of these substances, show their building blocks, amino acids and nucleotides, their links and relative position, will speak about bond angles and distances and the various helices formed. But, if we ask why Nature has put together that very great number of atoms in that very specific way, what property did she want to achieve, our biochemist will become silent. One of the basic principles of life is "organization" by which we mean that if two things are put together something new is born, the qualities of which are not additive and cannot be expressed in terms of the qualities of the constituents. This is true for the whole gamut of organization, for putting electrons and nuclei together to form atoms, atoms to molecules, amino acids to peptides, peptides to proteins, proteins and nucleic acid to nucleoproteins, etc. What Nature had in mind when doing this we cannot even guess at present. So here, too, we find the door to the central problem locked.

There are various circumstances which make this situation rather disturbing. First, these unanswered questions are the central and most intriguing problems of biology. Another rather disturbing fact is that corresponding to lacunas in our basic knowledge, there are lacunas in medical science and a great number of "endogenous" or "degenerative" diseases still rampage freely, causing endless suffering. But the most disturbing fact is that while biochemistry is still progressing in the fields where it has already been successful, it makes practically no progress in solving the problems mentioned. It looks as if the problems of biology could be divided into two classes: those which current biochemistry can solve and those which it cannot. It looks as if something very important, a whole dimension, might be missing from our present thinking without which these problems cannot be approached.

There is no doubt in the author's mind what this missing dimension is. The story is simple and logical. Biochemistry came into bloom at the end of the last century. At that time, matter was thought to be built of very small, indivisible units, atoms. Molecules were the aggregates of these atoms. There were about 90 different sorts of atoms which were symbolized by various letters, while their links were denoted by dashes. No doubt, this letter-and-dash language ranks among the greatest achievements of the human mind and is responsible for all the amazing successes of biochemistry. If we go through the list of problems enumerated above, we will find that the ones with which biochemistry was successful were problems of structure, or changes of structure taking place in simpler reactions which could be duplicated mostly in homogeneous solutions, and could be expressed and answered in terms of letters and dashes, while the problems which remained unanswered were problems of function of complex systems which cannot be expressed in this language. How could a reaction such as muscle contraction, the main product of which is not a substance, but work, w, be expressed in these terms?

The language of current biochemistry is still that of letters and dashes which means that this science is still moving in the same molecular dimension as it was moving at its birth in the last century. But since that time its parent science, chemistry, allying itself with physics and mathematics, made a dive into a new dimension, that of the submolecular or subatomic dimension of electrons, a dimension the happenings of which can no longer be described in the terms of classic chemistry, the rules of which are dominated by quantum, or wave, mechanics. Looked at through the glasses of this new science the atom is no more an indivisible unit but consists of a nucleus surrounded by a cloud of electrons with varying and fantastic shapes, and it seems likely that the subtler phenomena of life consist of the changing shapes and distributions of these clouds.

Biochemistry did not follow its parent science, chemistry, into this new subatomic dimension, which may hold the key to the understanding of the subtle biological functions. An example may illustrate the point. On the left side of Fig. 1 stands the classic

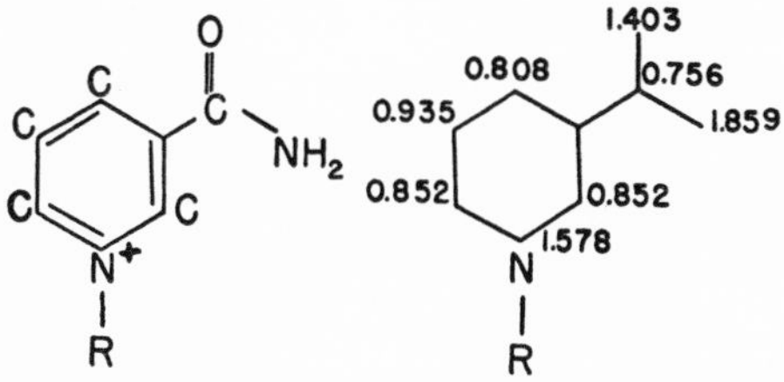

Fig. 1. Classic formula and molecular diagram of the pyridine end of DPN.

formula of the pyridine end of DPN expressed in classic symbols. It tells us that the pyridine ring is built of five equal C atoms and an N which has a positive charge. On the right side of the same figure is the "molecular diagram" of the same substance, as found

13

in a recent publication of the Pullmans.* The numbers coordinated to each atom indicate the electric charge. They tell us that each atom has a different charge and the molecule is thus surrounded by an electronic cloud of very complex structure. The positive charge is divided unequally over the one N and five C atoms of the ring while the negative charges are relegated to the side chain. This figure should be completed by three more sets of numbers, one set giving information about what is called the "free valency" of the single atoms of the ring, the other describing the "bond-order" of the links, and the third giving the "localization energies." While the classic formula attributed to the whole molecule but an overall shape and a dipole moment, in the molecular diagram every atom of the ring assumes a personality, a profile, a high degree of specificity and the whole structure begins to assume that subtlety which we can expect from any structure taking part in biological reactions.

While atoms and molecules were revealed to be complex little universes, their strict individuality has been broken down by "solid state physics." If many atoms form a regular and closely packed system, they may develop new properties. If, for instance, a great number of copper or iron atoms get together in a specific order they may develop electric conductivity, which is a collective property due to the interaction of the wave mechanical properties of the single units. Even macromolecules may develop solid state properties. So, in order to approach the central problems of biology we have to extend our thinking in two opposite directions, into both the sub- and supramolecu-

* Ref. 3, page 6.

14

lar. The two, in a way, are identical, the supramolecular qualities being but the collective action of the submolecular factors, supplying a new example of "organization." Similarly, we can expect entirely new properties to develop also when these molecules or molecular aggregates interact with the general matrix of life, water, forming with it a new and unique system. The elucidation of all these interrelations may eventually lend to our thinking the plasticity which may be necessary to approach life and the meaning of that unique system called the "cell."

The approach to these new dimensions may be a difficult one, and many of the ideas to be presented here may seem hazy and doubtful. The unknown offers an insecure foothold. What admits no doubt in my mind is that the Creator must have known a great deal of wave mechanics and solid state physics, and must have applied them. Certainly, He did not limit himself to the molecular level when shaping life just to make it simpler for the biochemist.

III

The Energy Cycle of Life

IN OUR FIRST APPROACH WE WILL DO WELL TO take a broad view. The broadest view we can take of energetics of life consists of considering the whole living world, trying to see how energy drives it. It is common knowledge that the ultimate source of this energy is the radiation of the sun. If a photon, ejected by the sun, interacts with a material particle on our globe, it lifts an electron from an electron pair in the ground state to a higher empty orbital, as symbolized by the upward pointing arrow in my Fig. 2,A. As a rule, the electron drops back within a very short time to its ground level, as symbolized by the downward pointing arrow. Life has shoved itself between these two processes and makes the electron drop back within its own machinery, utilizing its energy, as symbolized by the semicircle in Fig. 2,B. In order to do this efficiently it must meet the electron with a specially built substance (mostly chlorophyll) and couple this substance to a system which converts the very labile electronic excitation energy into a more stable chemical potential, into chemical energy, that is the energy of

17

a system of electrons of a stable substance. According to our present knowledge this is done, to a great extent, by using the energy of the excited electron to separate

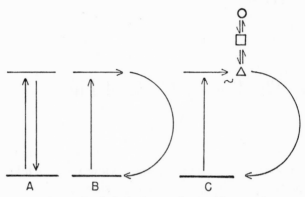

Fig. 2. Symbol of photosynthesis. (See text.)

the elements of water, H_2O.* The O is sent as O_2 into the atmosphere while the H becomes coupled to a pyridine nucleotide, TPN^+ or DPN^+, which thus becomes reduced into TPNH or DPNH, as shown by the work emanating from the laboratory of D. Arnon[12] and others. Simultaneously, also, ATP is formed from ADP and P, converting part of the energy into that of the terminal "~" of ATP. The pyridine nucleotide is symbolized in my Fig. 2,C, by a triangle, while the square and circle above it express the fact that pyridine nucleotides and ATP are unfit to store energy in quantity and so their energy may be converted into other forms more fit for storage. This is done by absorbing

* I will consider here only the classic "open" cycle of photosynthesis. For the simpler and "closed" forms see D. I. Arnon.[12]

[12] D. I. Arnon, *Nature*, **184**, 10, 1959.

CO_2 from the atmosphere and reducing it to carbohydrate and water, as symbolized by the left side of my Fig. 3, the square standing for carbohydrate, the circle for fat.

All this requires much time and a bulky apparatus, and so we let the plants do it and then eat the plant, or eat the cow which has eaten the plant. This, our eating, is symbolized in Fig. 3 by the horizontal arrows, while on the right side of the same figure I tried to symbolize, in a most sketchy and symbolic manner, what we do with these substances. We transfer the H atoms of the carbohydrate mostly back onto a pyridine nucleotide, releasing the C in the form of CO_2. Vennesland, Westheimer, and their associates,[13,14] by the isotope technique, have actually identified the H detached from metabolites with the H coupled onto the pyridine nucleotide. The DPNH or TPNH, in their turn, reduce flavine-nucleotides (FMN) but the H's found on $FMNH_2$ could no longer be identified with those of DPNH or TPNH and so it seems likely that what is transferred from these to FMN are not H's but are electrons, and the H's found on $FMNH_2$ in test tube experiments are derived from the universal solvent, water, from which the negatively charged FMN^- captured protons. Kosower et al.[15] have shown DPN to be a good electron donor.

DPN^+ needs for its reduction to DPNH one H and one electron. It can thus take up (and give off) both

[13] H. F. Fisher, E. E. Conn, B. Vennesland, and F. H. Westheimer, *J. Biol. Chem.* **202**, 687, 1953.
[14] F. A. Loewus, F. H. Westheimer, and B. Vennesland, *J. Am. Chem. Soc.* **75**, 5018, 1953.
[15] E. M. Kosower and P. E. Klinedinst, *J. Am. Chem. Soc.* **78**, 3493 and 3497, 1956.

H's and electrons; it is, so to speak, an exchange counter at which the H's of the foodstuffs can be exchanged for electrons, which then are sent down the oxidative chain over FMN. From FMN the electrons go from substance to substance, of which I quoted, in Fig. 3 as an example, cytochrome b, c, and a.* Eventually,

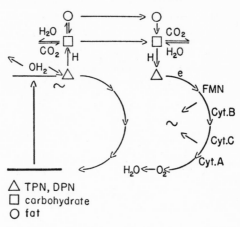

FIG. 3. Symbol of photosynthesis. (See text.)

the electron is taken up by the O_2 which then binds H ions and is thus reduced to H_2O. In H_2O the electron reaches its lowest energy level having given up its energy stepwise. The energy thus given up is converted with some loss into the energy of the $\sim$ of ATP.

* Electrons are sent down the chain also via succinate which mediates between this chain and the citric acid cycle. For simplicity's sake this has been omitted from the figures, as well as other intermediary catalysts, as the Q coenzyme and the free Fe which, according to D. E. Green[16] also may play an important role in electron transmission.

[16] D. E. Green, *Adv. Enzymol.* **21,** 73, 1959.

The point which I want to bring out is that apart from the process of excitation (vertical arrow), the right-hand side and left-hand side of my Fig. 3 are essentially identical. Being separated only by horizontal arrows which represent our (theoretically) uninteresting eating, we can pull the two sides together, as I have done in Fig. 4. Since carbohydrate and fat are

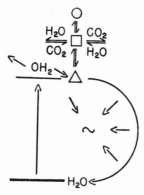

Fig. 4. Symbol of photosynthesis. (See text.)

but side lines, we can leave them off too, and complete the sketch with the main actor of this drama, the good old sun, as has been done in Fig. 5. In this figure we have thus the essentials of the energy cycle of life, which consists of electrons being boosted up by photons, and then dropping back to their ground level through the living systems, giving up gradually their excess energy which then drives the living machinery.

This Fig. 5 contains nothing beyond common knowledge. All the same, it helped me to clarify my mind about three essential points which form the cornerstones of my thinking. First, it made me see that life

21

is driven by nothing else but electrons, by the energy given off by these electrons while cascading down from the high level to which they have been boosted

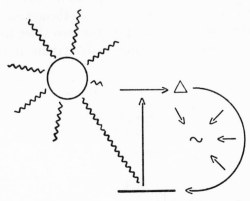

FIG. 5. Symbol of photosynthesis. (See text.)

up by photons. An electron going around is a little current. What drives life is thus a little electric current, kept up by the sunshine. All the complexities of intermediary metabolism are but lacework around this basic fact. The second point is that what is left behind by the electrons is but ATP and DPNH or TPNH. So these substances have to be the real fuel of life. My third point is that in this cycle the electrons go it alone. They are boosted up, one by one, and go through the cytochrome series, one by one, the central Fe atom being capable of a monovalent change only. It seems likely to me that they go alone throughout the whole cycle. I will finish this chapter by telling why these points are so important for me.

I have to go back to my school days when I was taught that life is driven by burning processes and the oxidative energy is released in reactions taking place between colliding molecules. If molecule X collides with molecule Y and oxidizes it, the free energy of the system is decreased by a ΔF which was supposed to drive life. X and Y, as other molecules, are closed systems and so is XY formed during their interaction. I could never see how a change taking place in a closed system can drive anything outside. It is like energy released in a closed box. So the ΔF remained for me but an item for my thermodynamic bookkeeping and not something with which one can buy something.* The dropping electron suggested by Fig. 5 is some-

Fig. 6. Symbol of oxidative phosphorylation. (See text.)

thing different: it is a little current by which I could drive anything, a dynamo or a muscle machine, or produce "~'s" by it. In order to produce ~ with it I would simply have to put a "black box"† (Fig. 6) between its two levels in question, into which box I would feed high-energy electrons, ADP, and P on top

* The thermodynamic bookkeeping is most important because it keeps us from talking nonsense. Our "debit" and "credit" accounts must tally and the final figures must conform to the two basic laws, the first of which tells that you cannot get something for nothing, and you cannot overdraw your banking account, while the second warns you that there is an overhead charge for any transaction. So thermodynamics only tells us whether a reaction is possible or not but tells us nothing about its nature or mechanism.

† By "black box" is meant, according to current usage, some unknown reaction which one hopes to clarify later.

to pull out ATP and a tired electron at the bottom. In order to drive life later with my ATP I would only have to turn the box upside down, feed in ATP at one end to pull out a high-energy electron (and ADP + P) at the other. So the question, how ATP can drive life, can produce contraction in muscle, narrowed down the problem for me, for instance, to how the energy of its $\sim$ can be transformed into electronic energy (see Chapter X).

The electrons "going alone" also need some additional explanation. At school I was taught that organic reducing agents are substances which can give off two H's or two electrons while oxidizing agents can take up two H's or electrons, have thus two stable states which differ from one another by two H's or two electrons.* To quote Michaelis and Schubert[17]: "Usually, in organic compounds, reduction is bivalent, that is it involves taking up two electrons. This is because almost without exception, the compounds which we have considered as stable organic molecules contain an even number of electrons."

What matters here is that in such a bivalent redox process the two molecules, oxidant and reductant, meet, assume a new stable configuration after having lost or gained two electrons, then part. I was unable to see how the ΔF of such an encounter with the subsequent internal rearrangement of molecular structure

* H's and electrons are equivalent since a negative charge can easily be exchanged for an H by capturing a proton, and vice versa: $A^{2-} + 2H^+ \rightleftharpoons AH_2$.

[17] L. Michaelis and M. P. Schubert, *Chem. Rev.* **22**, 437, 1938. See also ref. 11, on page 7.

could drive anything. This situation was not changed by the discovery of monovalent oxidation by Michaelis. To use his words: "all oxidations of organic molecules, *although they are bivalent,* proceed in two successive univalent steps, the intermediate state being a free radical.[18,19] In Michaelis' idea the situation is thus analogous to that in Noah's ark where the animals had to go two-by-two. Michaelis allows them only to pass the bridge one-by-one.

The distinction is important. A bivalent oxidoreduction is a classic chemical reaction which involves a rearrangement of molecular structure while a monovalent electron transfer, an electron going it alone, is a little current which does not necessarily involve such rearrangement. The sideline in Fig. 4 is what is usually summed up as "intermediary metabolism" and is composed of a host of classic chemical reactions, and will not occupy us further; but what goes around in the semicircle and is driving life is a current, single electrons cascading down and giving up their energy piecemeal. A current can do anything but cannot be expressed in classic chemical terms. A wandering electron belongs to the world of changing shapes and

[18] L. Michaelis, The Theory of Oxidation and Reduction, *in* "The Enzymes," J. B. Sumner and K. Myrbäck (Eds.), Vol. II, Part I, p. 1. Academic Press, New York, 1951. See also ref. 11 on page 7.

[19] W. H. Westheimer (in the "Mechanism of Enzyme Action," W. D. McElroy and B. Glass (Eds.), page 321, Johns Hopkins Press, Baltimore, Md. 1954) has challenged the universal applicability of this statement and has given examples in which oxidations may occur without passing through the free radical intermediate. So, according to his views, the word "all" would have to be replaced by "many" in the above quotation.

distributions of those electron clouds which belong to the submolecular, dominated by quantum mechanics.*

* This distinction between classic chemical reactions and monovalent electron transfer is not invalidated by the fact that in the last analysis also a classic chemical reaction is but the end product of a series of quantum changes. Nor is the distinction invalidated by the fact ATP is also formed in the intermediary metabolism.

IV

Units and Measures

IN ORDER TO DISCUSS ENERGY CHANGES, CORRELATE facts, make numerical statements or predictions, we need units and measures. This chapter will be devoted to a brief discussion of the main possibilities available at present.

REDOX POTENTIALS

A transmission of electrons from one substance to another means the oxidation of the one and the reduction of the other. We can also place the solution of reductant and oxidant into two different beakers and make the electrons pass from the one to the other through a wire. The "stronger" a reducing agent, the more it will tend to give off electrons and charge up the electrode, while the opposite will hold true for the oxidant. The potential difference between the two solutions will give us information about the free energy change of the oxidoreduction which would take place between the two substances if we were to mix their solutions, one EV (electron volt) being equal to about 23 kcal.*

The redox potentials can give us a great deal of most useful information. They created order among the host of oxidizing and reducing agents, allowing us to arrange them in a row according to their oxidizing or reducing power. They also allow us to make numerical statements about the energy changes taking place in the oxidation cycle. The potential difference between DPNH and O_2, which corresponds to the semicircle in Figs. 2–5, is about 1.1 EV, 25 kcal, a rather modest amount. Life, from the energy point of view, is a very modest phenomenon indeed. But even this small amount of 25 kcal seems to be turned in at the first step for small change, the energy of the $\sim$ of ATP which is about 10 kcal.† Maybe, the subtle structures which make the living machinery cannot be exposed to the destructive action of higher quanta.

These values put limitations on our speculations about bioenergetics. There are many reasons to tempt one to involve excited states of molecules in biological oxidation, but the energies needed to raise an electron from the ground state into the first excited level in one and the same molecule are mostly considerably higher than 25 kcal. For example, to lift an electron

* It is customary to measure the potential of the two substances in question one by one against some standard electrode, as the normal H electrode. Since limiting conditions are not well defined, one usually does not measure the potential of a pure oxidizing or reducing agent but measures the potential of an isomolar mixture of its oxidized and reduced form.

† The smaller values found experimentally 6000–8000 cal, relate to standard conditions, equimolar mixtures of ATP and ADP. The ADP concentration in tissues is negligible and so the free energy of ATP is correspondingly higher.

in DPN from the ground state into the first excited level of the same molecule, we would need a quantum of about 85 kcal (these substances having their absorption in the UV). No one has yet demonstrated the production of such a high-energy quanta in oxidative metabolism. Even a substance absorbing at the red end of the spectrum would demand 40 kcal for its excitation. Such transitions can be produced by absorbed photons. So we have to distinguish between processes induced by photons, like photosynthesis or vision, and processes occurring in the normal metabolism. This book will be concerned only with the latter.

It is true, the firefly may object to such a distinction, demonstrating by its greenish light that it is capable of producing quanta of 60 kcal, and one could think that bioluminescence may not be an isolated phenomenon but merely a leak in a metabolic process. In fact, there are even reports about light emitted by germinating seeds.[20] All the same, this does not necessarily mean that electrons have been excited from the ground state to the excited level within one and the same molecule. To this point I will come back later; meanwhile, I will maintain the distinction.

Like any method, the measurement of redox potentials also has its shortcomings. First, it gives us only thermodynamic data which is to say that the potential measured will not tell us how fast a reaction will go. There are strong reducing agents, e.g. agents containing SH groups, which will react very sluggishly with the electrode, or not react at all. Second, the standard conditions applied in our experiment are not identical

---

[20] L. Colli, U. Facchini, G. Guidotti, R. Dugnani, M. Orsenigo, and O. Sommariva, *Experientia,* **11,** 479, 1955.

with the conditions in the living cell. Third, the energy changes may be different according to the dielectric constants of the medium. In our measurements *in vitro* we usually use water as a solvent. D. E. Green[21] has always strongly emphasized that the matrix of the oxidation cycle within mitochondria consists of lipid material, is hydrophobic, and energy values may be quite different in such a medium.

Another more serious limitation is due to the fact that only the potential of substances can be measured which can freely give up electrons, that is, have two stable states which differ from one another, as a rule, by two electrons. As will be shown later, there are reasons to believe that there are organic substances which play an important role in biology and are capable of giving off one electron only and have no stable state which would correspond to such a one-electron transfer. These substances will give no potential at all. This shortcoming is not relieved by the fact that free radicals, which can give or take up one electron, might give a potential and Michaelis and his associates[*] were capable of measuring the potential of various free radicals formed on the way to bivalent oxidations at unphysiological pH's.

If we measure the free energy change of a redox process then we measure not only the energy change involved in transferring one or two electrons from the one substance to the other, but measure the total energy change which includes also the energy changes taking place in the rearrangements within the two interacting molecules, rearrangements which have to

[*] Ref. 17 on page 24 and ref. 18 on page 25.

[21] D. E. Green and R. L. Lester, *Fed. Proc.* **18**, 987, 1959.

accompany the transfer of two electrons and cannot do external work. Very often one would like to know the free energy change which corresponds solely to the transfer of an electron, uncomplicated by consecutive structural changes.

IONIZATION POTENTIALS AND ELECTRON AFFINITIES

If an electron passes from one molecule to another, from a donor (D) to an acceptor (A), energy may be gained or lost. In order to evaluate the change in our thoughts (and in our thoughts only) we can make the electron pass over a detour, in two moves. In the first imaginary move we take the electron from D, altogether removing it into infinity, while in the second move we drop it from infinity onto A. This procedure has the advantage that we have a fixed point: the energy of the electron in infinity. In infinity all electrons can be supposed to have the same energy.

This mental experiment is symbolized in Fig. 7A in which we take the electron from the molecule D and then drop it onto one of the A's. In this figure the thick lines stand for the highest filled energy level, the "ground state," occupied by an electron pair. If we remove an electron from an atom or molecule it is, as a rule, this level from which we remove it. In order to remove it we have to impart to it sufficient energy to be able to move through all the empty levels (symbolized by thinner lines), beyond the last one of which lies infinity. This energy, symbolized by the upward arrow, has close relations to what is called "ionization potential."[22] The higher the ground state lies, the shorter the arrow and the less energy will be needed

[22] R. S. Mulliken, *Phys. Rev.* **74**, 736, 1948.

31

to remove the electron, the smaller its ionization potential, the greater the tendency of the molecule or atom to "donate" an electron, the stronger donor it will be.

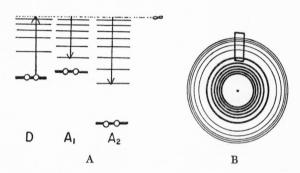

Fig. 7A. Symbol of the ionization potential and electron affinity. (See text.)

Fig. 7B. Schematic representation of energy levels of an atom. Thick lines: occupied levels. Thin lines: empty levels. (See text.)

To make such symbols more intelligible we could suppose that the atom is surrounded by a number of spherical filled and unfilled orbitals. In Fig. 7B the filled ones are symbolized by thick, the empty ones with thin, lines. It is the lower levels which are filled and the higher ones which are empty. The lines in Fig. 7A would correspond to the section of the orbitals enclosed by the oblong. In most other figures, similar to 7A, only the highest filled and lowest empty orbitals will be given. The symbols will also hold for the π electrons of more complex molecules, which π electrons are non-localized and belong to the whole molecule, or at least to its system of conjugated double bonds.

The situation will be more complicated in our second move in which we drop the electron onto the acceptor in which the electron will have to occupy the lowest empty energy level. The energy released in this act is symbolized by the downward arrow and is called the "electron affinity" (EA). What complicates the situation here is the fact that by adding one electron we disturb the postion of all energy levels, and so the energy of the orbital which this added electron eventually occupies will not be identical with the energy of the lowest energy level which existed prior to our manipulation. The lower the new level lies, the greater the EA (the longer the downward arrow), the more energy will be gained in this move and the more the atom or molecule will tend to accept an electron, the stronger acceptor it will be.

If there were no other factors involved, we could predict the energy change accompanying the electron transfer from the difference in length of the upward and downward arrows, EA–IP. More energy would thus be gained in the transfer of an electron from D to A_2 than to A_1, the EA of A_2 being greater, its arrow longer. That the change in energy will not be equal to EA–IP is due to the fact that there are additional complications. Such a complication is introduced by the fact that the transfer of the electron may alter also the interrelation of the D and A molecules, as well as their relation to the solvent. As will be discussed later, D and A must be very close to one another if an electron has to pass between the two. This implies that there can hardly be any solvent molecules between them, there will be solvent molecules only outside the DA complex. It is easy to see that polar solvent molecules, like those of water, with their strong dipole character, will exert a strong attraction on the electron transferred, will tune down the electropolar attraction between the two interacting molecules and may even enable the two molecules to part, each with its unpaired electron, as a "free radical." While the two molecules are in close proximity the transferred electron may also resonate between the two, contributing with its resonance energy to the balance of forces. Summing up, we can thus say that the change in energy will not be equal to EA–IP. If we lump all the other factors as Δ, we could say that the change in energy accompanying the transfer of an electron will be equal to EA–IP + Δ.

In spite of all these complications IP and EA are most useful parameters. In the experiment we can often simplify the complex situation. We can, for instance, as we will discuss later, couple one and the same acceptor to a number of various donors, using always similar solvents. In this case, the only variable will be the quality of the donors, and so the situation will be dominated solely by their IP. We can also reverse the situation and use the same donor with different acceptors to make the EA of the acceptors come to the fore.

Unfortunately, the ionization potentials are but poorly known, and the electron affinities are hardly known at all. There are various methods for the measurement of the ionization potentials, which give fairly consistent results in themselves, but the results obtained by the different methods may differ from one another by as much as one EV, which is about as much as the total energy released in the whole biological oxidation.

Moreover, distinction should be made between ionization potentials measured by fast and slow methods. The electron can, for instance, be torn off by a light quantum within 10^{-15} seconds, which time is too short to allow any changes in the molecular configuration and would thus reveal purely the energy used in this act of removing an electron. Other methods are slower than this optical method and may thus yield complex quantities composed of the energy of tearing the electron off *and* the energy changes due to the changes in molecular configuration accompanying the removal of the electron. The ionization potentials given in tables relate mostly to measurements in the gas phase. In biology we are concerned with condensed phases, mostly watery solutions in which the ionization energies may be quite different.

But even if ionization potentials and electron affinities give us, at present, but modest help in our work, they can provide a backbone for our thinking, presenting a clean-cut case of the removal and addition of an electron, into which we can, at least in thoughts, decompose the transfer of an electron from one substance to another.

ORBITAL ENERGIES

Two methods are commonly used for the calculation of energy levels of molecules. One is the molecular orbital method in the LCAO approximation (Linear Combination of Atomic Orbitals), the other the valence bond method. For calculations involving but the simplest molecules one is restricted to the LCAO method which has been applied extensively by B. and A. Pullman and G. Karreman to calculate the energy levels in a great number of molecules taking part in different biological reactions.* These calculations are very involved and demand not only a heavy mathe-

* See quotations in **Table I.**

matical armory, but also much experience and intuition. The calculations are restricted to the case of molecular orbitals, that is, to cases where the molecule has a system of delocalized electrons, as is the case, for instance, in systems of conjugated double bonds where the electron in question does not belong to any single atom but to the whole molecule, is a π electron belonging to the "π electron pool." This may give the impression that only molecules with extensive conjugated systems with an extensive π pool are important for electron transfer in biology. Certainly, most of the catalysts of biological oxidations have such systems, but we must not allow ourselves to be deceived into believing that only such molecules can be important as electron donors or acceptors. If we talk mostly about them, this is only because we know most about them.

It follows from what has been said before, that we are interested chiefly in the energies of the highest filled and lowest empty orbital.

The energy of an orbital, that is, the energy of the electron on that orbital $E = \alpha + k\beta$, where α is the "Coulomb integral" of the method, and β the "exchange integral" between two C's. For substances belonging to the same chemical series α and β are fairly constant, and so the energy depends on k. Where $k = 0$, $E = \alpha$. This is the zero. As a rule the filled "bonding" levels lie below 0 and as a convention their k has a positive sign. The empty "antibonding" levels lie above 0, their k being denoted with a negative sign. Figure 8 shows a number of k values. The values of k of a greater number of substances, as calculated by the Pullmans and G. Karreman, are given in Table I.

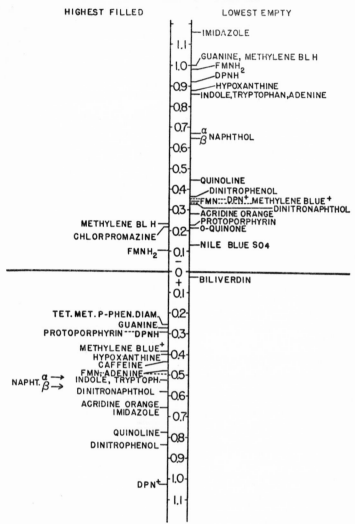

Fig. 8. *k* Values of the highest filled and lowest empty molecular orbitals.

TABLE I
k VALUE OF THE HIGHEST OCCUPIED (LEFT), AND LOWEST EMPTY (RIGHT) MOLECULAR ORBITAL

Substance	k Values		Reference [a]
Acridine	0.494	−0.342	(9)
Acridine orange	0.657	−0.278	(9)
Adenine	0.486	−0.865	(3)
Alloxane	1.033	−1.295	(3)
6-Aminonictonamide	0.735	−0.471	(8)
2-Amino-4-nitrophenol	0.469	−0.354	(10)
4-Amino-2-nitrophenol	0.451	−0.355	(10)
Aniline	0.544	−1.000	(10)
Anthracene	0.414	−0.414	(2)
Antipyrine	0.248	−0.956	(10)
Ascorbic acid	0.529	−0.899	(10)
Atabrine	0.311	−0.486	(10)
Barbituric acid	1.033	−1.295	(3)
Benzanthracene	0.452	−0.452	(2)
1-Benzyl-2-methoxy-(N,N-dimethyl) tryptamine	0.427	−0.866	(6)
Benzpyrene	0.372		(1)
Biliverdine	0.455	+0.021	(5)
Bromanil	0.646	−0.266	(10)
Catechol (Adrenaline)	0.666	−1.049	(10)
Chloranil	0.753	−0.275	(10)
Chlorpromazine	−0.217	−1.000	(6)
Chloroquine	0.478	−0.654	(10)
Chrysene	0.521	−0.520	(2)
Colchicine	0.355	−0.499	(10)
Cytosine	0.595	−0.795	(3)
5-Methylcytosine	0.530	−0.796	(3)
Dibenzanthracene	0.494	−0.501	(2)
2,4-Dichlorophenol	0.698	−1.016	(10)
3,5-Dichlorophenol	0.749	−1.032	(10)
5,7-Dimethyl-3,4-benzacridine radical	−0.277	−0.716	(10)
5,7-Dimethyl-1,2-benzacridine radical	−0.299	−0.682	(10)
9,10-Dimethyl-1,2-benzanthracene	0.387	−0.475	(10)

[a] References for this table are listed at the end of the table.

TABLE I (*Continued*)

Substance	k Values		Reference
p-Dinitrobenzene	1.000	−0.232	(10)
m-Dinitrobenzene	1.015	−0.317	(10)
2,4-Dinitronaphthol	0.580	−0.329	(10)
2,4-Dinitrophenol	0.841	−0.352	(9)
2,5-Dinitrophenol	0.809	−0.243	(10)
DPN+	1.032	−0.356	(2)
DPNH	0.298	−1.032	(2)
Duroquinone	0.757	−0.273	(10)
Flavinemononucleotide	0.496	−0.343	(4)
$FMNH_2$	−0.105	−0.979	(4)
Fluoranil	0.960	−0.200	(10)
Fluoropromazine	−0.207	−0.987	(10)
Formilhydrazine	0.192	−1.710	(10)
Guanine	0.307	−1.050	(3)
1-Methylguanine	0.303	−1.064	(3)
9-Methylguanine	0.302	−1.074	(3)
Histidine	0.660	−1.160	(3)
p-Hydroquinone	1.000	−1.175	(10)
Hypoxanthine	0.402	−0.882	(3)
Indole	0.534	−0.863	(2)
Indole acetic acid	0.479	−0.863	(10)
Indophenole	0.516	−0.122	(10)
D-Lysergic acid diethyl amide	0.218	−0.726	(6)
Lumichrome	0.581	−0.434	(10)
Lumiflavine	0.482	−0.342	(10)
20-Methylcholanthrene	0.388	−0.475	(10)
Methylene blue+	0.398	−0.354	(7)
Methylene blue H	−0.232	−1.000	(7)
L-Methylmedmaine	0.348	−0.869	(6)
Methylphloroglucinol	0.649	−1.123	(10)
Naphthacene	0.295	−0.337	(2)
Naphthalene	0.617	−0.618	(2)
α-Naphthol	0.519	−0.671	(9)
β-Naphthol	0.569	−0.637	(9)
1,4-Naphthoquinone	1.000	−0.325	(10)
1,2-Naphthoquinone	1.000	−0.332	(10)
Nile blue sulfate+	0.591	−0.133	(10)
Nile blue sulfate H	0.091	−0.741	(10)
Nitrobenzene	1.000	−0.334	(10)
o-Nitrophenol	0.806	−0.352	(10)

39

TABLE I (*Continued*)

Substance	k Values		Reference
m-Nitrophenol	0.797	−0.334	(10)
p-Nitrophenol	0.827	−0.354	(10)
Phenanthrene	0.607	−0.606	(2)
Phenanthrenequinone	0.721	−0.442	(10)
o-Phenanthroline	0.649	−0.563	(10)
Phenazine	0.555	−0.251	(10)
Phenothiazine	−0.210	−1.000	(10)
Phenylalanine	0.908	−0.993	(3)
p-Phenylene diamine	0.321	−1.000	(10)
Phlorizine	0.731	−0.901	(10)
γ-Picoline	1.000	−0.872	(10)
Picramic acid	0.469	−0.354	(10)
Proflavine	0.657	−0.278	(10)
Protoporphyrin	0.293	−0.233	(5)
Pteroilglutamic acid (Folic acid)	0.52	−0.64	(3)
2,4-Dimethyl-pteridine	0.544	−0.508	(3)
2,3-Dihydroxypteridine	0.653	−0.663	(3)
2-Amino-4-hydroxy-pteridine	0.489	−0.650	(3)
Pteridine	0.864	−0.386	(3)
Pyocyanine (alkal)	0.286	−0.272	(10)
Pyrene	0.445	−0.444	(2)
Pyridine	1.000	−0.871	(9)
Pyrimidine	1.063	−0.820	(10)
Quinine	0.584	−0.563	(10)
Quinoline	0.77	−0.44	(9)
o-Quinone	0.694	−0.211	(10)
p-Quinone	1.000	−0.225	(10)
Reserpine (Indole part)	0.464	−0.892	(10)
Reserpine	0.627	−0.872	(10)
Rhodamine 5G	0.688	−0.184	(10)
Serotonin	0.461	−0.870	(10)
Stilbene	0.503	−0.503	(2)
Stilbestrol	0.369	−0.553	(10)
p-Terphenyl	0.594	−0.592	(2)
Tetramethyl-para-phenylenediamine	0.266	−1.000	(10)
Thionine+	0.398	−0.354	(10)
Thionine H	−0.208	−1.000	(10)
Thymine	0.510	−0.958	(3)
Toluidine blue+	0.391	−0.359	(10)

TABLE I (*Continued*)

Substance	k Values		Reference
Toluidine blue H	−0.217	−0.997	(10)
Trinitrobenzene	1.045	−0.317	(10)
Trinitrophenol	0.859	−0.317	(10)
Triphenylene	0.684	−0.655	(2)
Tryptophan	0.534	−0.863	(3)
Tyrosine	0.792	−1.000	(3)
Uracil	0.597	−0.960	(3)
Uric acid	0.172	−1.194	(3)
1-Methyl uric acid	0.172	−1.202	(3)
3-Methyl uric acid	0.153	−1.204	(3)
7-Methyl uric acid	0.133	−1.120	(3)
9-Methyl uric acid	0.161	−1.204	(3)
Xanthine	0.442	−1.005	(3)
Xanthine H on N9	0.397	−1.097	(3)
1-Methylxanthine	0.397	−1.198	(3)
3-Methylxanthine	0.345	−1.197	(3)
9-Methylxanthine	0.394	−1.213	(3)
Vitamin K_3 (2 methyl-1,4-naphthoquinone)	0.915	−0.340	(10)
Vitamin K_5 (4 amino- 2 methyl-1-naphthoquinone)	0.283	−0.738	(10)

REFERENCES FOR TABLE I

[1] A. and B. Pullman, Cancerization par les substances chimiques et structure moleculaires. Masson Edit. Paris, 1955.

[2] B. Pullman and A. Pullman, Les théories électroniques de la chimie organique. Masson Edit. Paris, 1952. In this book m values are given instead of k. The k values have been calculated by means of the following formulas: for filled orbitals $k = m/(1 − 0.25m)$ for empty orbitals $k = −m/(1 + 0.25m)$.

[3] B. and A. Pullman, *Proc. Natl. Acad. Sci. U.S.* **44**, 1197, 1958.

[4] B. and A. Pullman, *Proc. Natl. Acad. Sci. U.S.* **45**, 136, 1959.

[5] B. Pullman and A. M. Perault, *Proc. Natl. Acad. Sci. U.S.* **45**, 1476, 1959.

[6] G. Karreman, I. Isenberg, and A. Szent-Györgyi, *Science* **130**, 1191, 1959.

[7] B. and A. Pullman, *Biochim. et Biophys. Acta* **35**, 535, 1959.

[8] B. and A. Pullman, *Cancer Research* **19**, 337, 1959.

[9] B. Pullman, Personal communication.

[10] G. Karreman, Personal communication.

We need not concern ourselves here with the exact meaning of α and β. What concerns us here is that the k for the highest filled orbital is a linear function of the ionization potential.[*] The smaller its value the less energy will be needed to take an electron off, the easier the molecule in question will give up this electron and act as an electron donor. Similarly—as experience indicates—the smaller the $-k$ of the lowest empty orbital, the easier the molecule will accept an electron. A $k = 0.5$ for the highest filled orbital means, for instance, that the substance is a "fair" electron donor, the same $-k$ for the lowest empty orbital means that it is a "fair" electron acceptor; 0.2 means "very good." In exceptional cases the k of the highest filled orbital can be even smaller than 0, and have a negative sign, an "antibonding character." This means an excessively good donor. Such substances, of which few examples are known, are mostly unstable and readily auto-oxidize in air, as leucomethylene blue or $FMNH_2$. *Mutatis mutandis*, the same holds for positive values of the lowest empty orbital. The numerical value of β is taken as 1–3 EV, so a difference of 0.5 between two k values means 0.5–1.5 EV difference in the ionization potential for substances belonging to the same series of substances and having, thus, a similar α and β.

We must be very careful in relating the k's to other physical constants. While Mulliken has shown that the position of the highest filled orbital (given by the $+k$'s) actually gives the ionization potential of the molecule, the position of the lowest empty orbital (given by the $-k$'s) does not have such simple relationship to the electron affinity. The reason for this is

[*] Ref. 22 on page 31.

that, as pointed out before, upon adding an electron to the molecule, all the energy levels tend to change. This change, however, may be small in big molecules with an extensive system of π electrons, and most biological catalysts have such systems. What we may expect in any case is that, on the whole, the lower the position of the empty orbital (and the lower the $-k$'s) the greater will be the electron affinity, and so even the $-k$'s give a very useful parameter.

There are various ways to convince oneself of the approximate usefulness and reliability of these k's. We can, for instance, compare in various substances, the k's with their tendency to give off or accept electrons. The experience of my laboratory on this line supports, on the whole, the reliability of the k values. Another crude way of checking consists of measuring the absorption spectra. The longest wavelength of the light absorbed by a substance can be expected to lift an electron from the highest filled orbital to the lowest empty one, so there should be some relationship between the wavelength of the absorption and the distance between the $+$ and $-k$ values.[23] Fujimori's curve (see Fig. 11) also contributes to the mass of evidence pleading for the reliability of these k's.

The meaning of the k's has given me an answer to a problem which occupied my mind in earlier days and lured me into this field: why so many biological substances, like flavines, flavones, pteridines, and cyto-

---

[23] I. Isenberg and the author actually demonstrated such an interrelation (*Proc. Natl. Acad. U.S.* **45**, 519, 1959). It should be noted that in their table two absorptions were quoted erroneously (tyrosine and phenylalanine). These two amino acids would not fit if quoted correctly.

chromes, apparently involved in energetics, are colored? What is the meaning of color in cells not exposed to light? The answer is simple. In terms of k, color means that the distance between the two k values, that of the highest filled and lowest empty orbital, are close to one another, making even the energy-poor visible light capable of lifting an electron from the former to the latter. This light being thus absorbed will make the substance colored. In order to be able to act as a catalytic electron transmitter the substance in question has to be both a good electron donor and acceptor, that is, the k values for both the highest filled and lowest empty orbitals must be small, close to 0 and thus close also to one another. The k values of FMN, for instance, are 0.496 and −0.343 respectively, which makes FMN colored* and makes it into a good donor and acceptor. As I have shown in my earlier days,[24] one can knock out the whole respiratory chain by cyanide and then restore oxygen uptake by adding methylene blue which takes the whole electron transport over between dehydrogenases and O_2. This it can do, as pointed out by B. and A. Pullman,[25] because its k values are 0.398 and −0.354 respectively, almost ideally placed to make the dye into a good donor *and* acceptor.*

These examples may suffice to show the k values of

* One may object that DPN has no color and all the same it is one of the main mediators of oxidation. But the contradiction is only an apparent one because DPN is not a pure electronic mediator. It takes up H's and gives off electrons. Its k values are 1.032 and −0.356 which indicates that DPN+ is a very good acceptor. The donor will be not DPN+ but DPNH, the highest filled orbital of which has a k of 0.248 making it into a good donor.

[24] A. Szent-Györgyi, *Biochem. Z.* **150**, 195, 1924.
[25] B. and A. Pullman, *Biochim. et. Biophys. Acta*, **35**, 535, 1959.

the LCAO approximation to be most valuable tools of research and understanding. They also have their shortcomings. First, the LCAO method is but an "approximation" and the k values obtained depend on the parameters used in the calculations which may be subject to change. As mentioned before, experience and intuition are needed in this line which also means that there is no real numerical rigidity. Another, even more serious, shortcoming is that the k values of the substances belonging to different chemical series cannot be strictly compared, since both α and β may vary in different chemical series by as much as 25%. If the numerical value of α is around 6 EV, 25% means 1.5 EV, more than the total energy of biological oxidation.

To sum up: all three methods of measurement and expressions, redox potentials, ionization potentials and electron affinities, as well as orbital energies, have their merits and shortcomings and at present there is no universally applicable method available. What would be needed is to bring the three methods to a common denominator which has all the merits and none of the shortcomings. This, at present, is but a desire. The relations between redox potentials and the k values of the LCAO approximation are very complex, and there is no theory available to bring the two together. The relation of the ionization potentials and the k values are clearer, but the two cannot be identified since the α and β values are not constant for different chemical series. What would serve our purpose best would be a knowledge of the ionization potentials and electron affinities of all molecules in biological media, but this information is not available. So we have to wait for further development and, meanwhile, get along as well as we can.

45

V

Electronic Mobility

IN ORDER TO GET AROUND IN THE OXIDATIVE CYCLE, the electron must wander from substance to substance, and thus have a certain mobility. We do not know how this mobility is acquired. What we do know is that several members of the oxidation cycle are bound to structure, thus fixed in space, so that they cannot reach one another by diffusion. They are rather bulky and have relatively small active centers so that it seems unlikely that they could be arranged in such a fashion that their active centers touch, even if a small degree of freedom of motion (e.g. free rotation) is granted. All the same, there must be some connection between them. There are different ways known in which energy quanta or electrons can acquire a long-range mobility. But, tentatively, we could also suppose that in biological oxidation the electrons are carried from one structure-bound substance to another by small diffusible molecules which shuttle between fixed members of the oxidation chain, getting alternately oxidized and reduced. In certain cases we can exclude diffusion, as, for instance, between chlorophyll

47

and cytochrome in photosynthetic bacteria where Chance and Nishimura[26] found electron transmission even at liquid N_2 temperature where everything is frozen stiff and no diffusion is possible. All the same, the fact stands that the oxidation system contains diffusible small molecules, as the coenzyme Q or members of vitamin K group,[27] or Fe atoms to which D. E. Green calls attention.[28] These substances could transport electrons between fixed members of the chain. Whether they do so, we do not know.

Being unable to decide between the various possibilities, I will give brief consideration to the most likely modes of long-range energy and electron transfer and in the end I will consider the question, how electrons can be transmitted from one substance to another at short range, in a direct contact, without wasting their energy.

ELECTROMAGNETIC COUPLING. RESONANCE TRANSFER OF ENERGY

If a molecule X is excited and there is a molecule Y not too far away, which molecule is capable of a similar excitation, then there is a chance that the excitation may suddenly die out in X and appear in Y. Something analogous can be demonstrated in two pendulums connected with a weak spring. The motion imparted to the one dies off after a time and is taken over by the other. An electronic excitation can be

[26] B. Chance and M. Nishimura, *Proc. Natl. Acad. Sci. U.S.* **46**, 19, 1960.
[27] A. I. Arnon, *Nature* **184**, 10, 1959.
[28] D. E. Green, *Radiation Research* Suppl. 1960 (In press). Brookhaven Conf. Bioenergetics.

looked upon as an oscillation, and the role of the spring may be fulfilled by the electromagnetic field.

Th. Förster[29] has formulated the rules of such an energy transmission and his formula allows us to calculate the distance at which the energy may be transmitted in this way. Karreman and Steele,[30] for instance, have calculated the critical distance through which excitation can be transmitted between aromatic amino acids in a protein and found it to be 17 Å, a distance big enough to allow energy transmission within one molecule or closely adjacent molecules. Förster has also shown that the transmission becomes the more probable the greater the overlap between the absorption spectra of the two molecules and is favored by a somewhat longer wavelength of absorption of the molecule which has to take the energy over.

As to the mechanism of this energy transfer, we have two ways to picture it. Förster, as well as earlier investigators (J. and F. Perrin), picture it as a sort of jump from X to Y. However—as pointed out by Z. Bay at the symposium held in Woods Hole in the summer of 1959—quantum electrodynamics can give a different explanation. If molecule X is surrounded by several Y's, the disappearing excitation in X may bring these Y's into a state which is neither a ground nor an excited state. Then, after a very short time the energy will collect again and suddenly declare itself in the excitation of one of the Y's.* In many cases, the cal-

* Ordinary language is most inadequate in describing such phenomena.

[29] Th. Förster, *Disc. Faraday Soc.* **27,** 1959. See also Th. Förster, "Fluoreszenz Organischer Verbindungen," Göttingen, 1951.

[30] G. Karreman and R. H. Steele, *Biochim. et Biophys. Acta* **25,** 280, 1957.

culation according to the two theories may lead to the same result, but in certain cases, if there is a "coherence," a cooperative action between the Y's, the chances of the transmission may be greater in the second theory, which may have a role in photosynthesis, where several hundred chlorophyll molecules collaborate in leading a photon to its site of action.

Semiconduction

In 1941 I published an article with the ambitious title "Towards a New Biochemistry." This article grew out of discussions with my young pupil K. Laki and suggested that energy, in living systems, may be transmitted by conduction bands. But, biochemistry did not move; the theory remained a dead duck. Theories which bear no fruit are useless. All the same, a few words may be said about semiconduction. The basic idea is simple. Atoms have single energy levels. If many atoms are close to one another and are placed in a regular array, their energy levels disturb one another and the disturbed levels join to form a continuous band which contains many levels and extends over the whole system. The conductance depends, in this case, on the number of electrons within the band. According to the Pauli principle only two electrons can have the same energy within the same level. So, if the number of electrons occupying the band is double the number of levels, i.e. atoms building the system, then the band is filled and no conduction can take place.

Metals owe their conductivity to their unfilled bands. There may be energy bands also in dielectrics which, then, owe their insulating property to the fact that the

50

band corresponding to the ground state is completely filled and the next higher band is completely empty and much energy is needed to raise electrons into it.

In intrinsic semiconductors, the distance to the first empty band is small so that even heat agitation may raise an electron into it from the nearest filled band, rendering it conductant.

The calculation of Evans and Gergely[31] strongly suggested that proteins actually have conduction bands. Later, Eley[32,33] and his associates demonstrated experimentally the existence of such conduction bands, but also showed that the distance between the filled and the first empty band is about 2–3 EV, a distance too great to be covered by the small biological quanta.

Conduction bands certainly exist in diverse biological systems. The experiments of Arnold and Sherwood,[34] for instance, leave little doubt that such bands do exist in dried chloroplasts which can store light which can be hunted out as such later, by heat, "electron traps" being present under the conduction bands.

That energy can move through proteins was suggested long ago by Bucher and Kaspers[35] who produced dissociation in the heme part of CO-myoglobin by quanta absorbed by the protein. Shore and Pardee[36]

[31] M. S. Evans and J. Gergely, *Biochim. et. Biophys. Acta.* **3,** 188, 1949.

[32] D. D. Eley, S. D. Parfitt, M. J. Perry, and D. H. Taysum, *Trans. Faraday Soc.* **49,** 79, 1953.

[33] M. H. Cardew and D. D. Eley, *Ibid.*, 1959.

[34] W. Arnold and H. K. Sherwood, *Proc. Natl. Acad. Sci. U.S.* **43,** 104, 1957.

[35] T. Bucher and J. Kaspers, *Biochim. et. Biophys. Acta* **1,** 21, 1947.

[36] V. G. Shore and A. B. Pardee, *Arch. Biochem. Biophys.* **62,** 355, 1956.

have seen light emitted by the dye in chromoproteins after excitation of the protein. Schubert's[37] experiments cast some doubt on Bucher and Kasper's results but at the same time supplied fresh evidence for such energy transfer. None of these experiments is quite conclusive in the sense that they do not allow us to decide definitely between resonance energy transfer and semiconduction. One of the most important results of electron microscopy is the demonstration of the wide occurrence of membranes and layered structures, as found in mitochondria, chloroplasts, visual rods, and in the protoplasm (Sjöstrand[38]). Even the particles found in extracts, such as microsomes, may be but fragments of such membranes, broken up in the process of extraction. Membranes, especially double membranes (as most biological membranes are), suggest by their ordered molecular structure semiconduction. As has been demonstrated lately by Tollin[39] in Calvin's laboratory, such layered structures acquire a high conductivity if the one layer donates electrons to the other.

There is also the possibility that filled energy bands may become conductant by donating electrons to some outside substance and thus become unsaturated and conductant. Also, the opposite may be true and an empty electron band may become conductant by accepting electrons from some outside substance. Accepted and/or donated electrons may thus enable a

[37] J. Schubert, *Arkiv Kemi.* **15**, 97, 1959.
[38] F. S. Sjöstrand, Fine Structure of Cytoplasm: the Organization of Membranous Layers, *in* "Biophysical Science." J. Wiley & Sons, New York, 1959.
[39] G. Tollin, *Radiation Research,* 1960. (AEC conference on Bioenergetics, Brookhaven, 1959). See also "Semiconductors," Chapter 15 by C. G. B. Garrett, Reinhold Press, 1959.

biological system to fulfill its biological role by semi-conduction, or else may inhibit or paralyze function by tapping off the conducting electrons or filling up the holes which rendered the system unsaturated. Some of the hormones or drugs might act this way (see later). Accepted and donated electrons may thus enable a system to work, may regulate its function, or else cause profound disturbance. Even a small change along this line may have far reaching consequences, as the damage done to the insulation of an electric wire may paralyze a whole electric network. Mason[40] suggested relations between such changes and malignant growth.

Electromagnetic coupling transfers energy, semiconduction transfers electrons. We must clearly distinguish between the transfer of energy, and that of electrons, even if the electrons carry energy with them, and this for several reasons. In order to transfer energy, we would have to excite an electron first from the ground level of a substance to its first excited level. The energy needed for this excitation corresponds to the distance between the ground levels and the first excited levels in Fig. 7. As mentioned before, this energy is of the order of 40–100 kcal, an energy quantum which may be carried by a photon, but which, probably, cannot be produced in the normal course of oxidation. Hence we must distinguish between processes started up by photons, as is the case in vision and photosynthesis, and processes occurring in metabolism. This book is concerned only with these latter.

Another reason for distinguishing between transfer of energy and electrons will become evident by casting

[40] R. Mason, *Nature,* **181,** 820, 1958.

another glance at Fig. 7, in which three molecules were symbolized which are supposed to belong to different substance groups and have different 0 levels. What is drawn here, at the same level, is infinity, ∞, since all electrons, in infinity, have the same energy. Infinity, in this case, lies just beyond the highest empty orbital. Judging from the length of the arrows (ionization potential in D and electron affinity in the A's) we could transfer no electron without external help from the ground level of D to the first excited level of A_1, but could transfer an electron from D to A_2. The reverse is true for energy. We could transfer excitation energy by resonance from D to A_1 since the distance from the ground level to the first excited level is smaller in A_1 than in D. However, we could not transmit excitation energy from D to A_2, since the distance between the two levels is greater in A_2.

CHARGE TRANSFER[41]

Evidence started accumulating more than thirty years ago suggesting that in certain complexes electrons may trespass between the borderlines of two complexing molecules. It was J. Weiss[42-44] who, in 1942, gave the first clear formulation of the idea that within a complex an electron of one of the two complexing molecules may be transferred to an orbital of the other. This "charge transfer" has been extensively

[41] A very clear, though not very recent, review of charge transfer, is found in the: *Quarterly Reviews of the Chemical Society, London* **8**, 422, 1954 by L. E. Orgel.

[42] J. Weiss, *J. Chem. Soc.* **245**, 1942.

[43] J. Weiss, *Nature* **147**, 512, 1941.

[44] J. Weiss, *Trans. Faraday Soc.* **37**, 78, 1941.

studied since by Mulliken[45-47] and his associates, who applied quantum mechanics to it and classified its various forms.

The transfer of electrons from one substance to another is usually termed as an oxidoreduction. However, we must clearly distinguish between "oxidoreduction" and "charge transfer." As a rule, in organic substances, electrons occupy orbitals in pairs and in oxidoreduction an electron pair is transferred from one molecule to the other, two new closed shell molecules being formed. The two molecules then part, the one having become richer, the other poorer, by two electrons. After this, in principle, they have nothing more to do with one another. So both molecules must have two stable states differing by two electrons. The whole redox process consists of the transfer of these two electrons and the consecutive rearrangement in structure. This situation is not changed by the fact that Michaelis[*,48-50] has shown that electrons can pass from one substance to the other, one-by-one, and that most[†] biological oxidations are actually built of two one-electron steps. The basic ideas of charge transfer not having been cleared yet, Michaelis, himself, regarded the one-electron step as an intermediary of the bi-

* Chapter I, ref. 11, page 7; Chapter III, refs. 17, page 24 and 18, page 25.
† Footnote 19, Chapter III, page 25.

45 R. S. Mulliken, *J. Am. Chem. Soc.* **72**, 600, 1950.
46 R. S. Mulliken, *J. Am. Chem. Soc.* **74**, 811, 1952.
47 R. S. Mulliken, *J. Phys. Chem.* **56**, 801, 1952.
48 L. Michaelis, *Chem. Rev.* **16**, 243, 1935.
49 L. Michaelis, M. P. Schubert, R. K. Reber, J. A. Kuck, and S. Granick, *J. Am. Chem. Soc.* **60**, 1678, 1938.
50 L. Michaelis and S. Granick, *J. Am. Chem. Soc.* **66**, 1023, 1944.

valent oxidation even if the free radicals formed in the one-electron step were stable at extreme pH's. In "charge transfer" one electron only is transferred. The two molecules, the "donor" and the "acceptor," usually stay together, and if they part they do not part as closed shell molecules, but as free radicals with an unpaired electron.

The establishment of the idea of charge transfer means a broad extension of our previous ideas. First, it breaks down the rigidity of our thinking about the individuality of molecules. Charge transfer means that the electrons of a molecule D (D for "donor") are capable of using, under certain conditions, orbitals of molecule A (A for "acceptor"). Second, it brings into the realm of electron transfer a host of substances capable of giving up one electron only, substances which do not have two stable states differing by two electrons, which do not affect the electrode, and are thus not regarded as oxidation or reduction agents. The transfer of one electron does not necessarily involve any rearrangement within the molecule. A charge transfer could be symbolized, schematically, by Figs. 9 and 10, as a simple transfer of one electron from the highest filled orbital of D to the lowest empty orbital of A without any further rearrangement.

Several important points are evident. In order to allow a passage of an electron from an orbital of D to an orbital of A the two electron clouds must overlap. This means that the two molecules must come very close to one another. Close fit and approachability become decisive factors. This may explain why many charge transfer reactions are very slow and take hours, the chances of coming together in the right way being remote. Naturally, this does not exclude that charge transfer may, under conditions, be achieved also in a short encounter, in a so-called "contact transfer." Another point which is made clear by Figs. 9 and 10 is

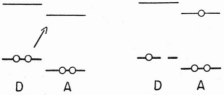

D   A   D   A

FIGS. 9 and 10. Symbol of the ground state and first excited level of two molecules, before (9) and after (10) charge transfer.

that the relative positions of the two orbitals on the energy scale, that is, the ionization potential of the donor and the electron affinity of the acceptor, have to become dominating factors. Schematically speaking, we could distinguish between two extreme cases: (1) the donating orbital lies considerably lower as compared with the accepting one, as is the case in Figs. 9 and 10, and, consequently, a considerable amount of energy is needed to lift the electron from the former to the latter, $IP \gg EA$; (2) the accepting orbital is lower than the donating one and so the electron can be transferred without outside help, $EA > IP$, as was the case between D and A_2 in Fig. 7.

Considering the first of these two cases, we would have to suppose that the two molecules approach one another and form a complex, held together by the classic forces of complex formation (dispersion forces, polarization, dipole moments). No considerable charge transfer could be expected. All the same, we know that in complex formation a dipole moment may be

57

developed and Mulliken attributes this to a small degree of charge transfer. The transferred electrons resonate between the two molecules contributing thus their "resonance energy" to the binding forces. As single molecules have excited states leading to an absorption, so these complexes also have an excited state, characteristic for the complex. In this case it is the energy of the absorbed photon which lifts the electron of the donor molecule to the excited level of the acceptor. In the excited complex the major part of the electron cloud is now on the acceptor molecule. It is for this reason that Mulliken calls such a spectrum a "charge transfer spectrum" which is characteristic for the complex. The energy needed for the charge transfer would be indicated by the wavelength of the absorbed light and would depend, in part, on the difference between the two levels, the donating and accepting one, that is, on EA–IP. If the difference is great the absorption spectrum falls into the UV. If the difference is small it may fall into the visible and mean a strongly colored complex. If the transfer would demand even less energy the spectrum would fall into the IR. In any case, the spectrum will be characteristic for the complex, and not for its components, as has been shown by Brackman[51]; it will be a "charge transfer spectrum."

Naturally, the IP of D and EA of A will not be the only factors in play because the transfer of an electron may alter the relation of the two molecules to one another, or may alter their relation to the solvent. As will be shown later, there are also other still unknown factors involved. All the same, if the IP of the donor

[51] W. Brackman, *Rec. Trav. Chim.* **68**, 147, 1949.

dominates behavior and all other factors are constant, then we can expect that the energy (that is the frequency) of the light absorbed will depend linearly on this IP. It follows that if we take a series of donors and couple them, one by one, to the same acceptor, then the frequency of the absorbed light, if plotted against the IP of the single donors, must give a straight line. This prediction was first made and verified in the experiment by McConnell, Ham, and Platt,[52] and simultaneously by Hastings, Franklin, Schiller, and Madsen.[53] The lines obtained in this plot were surprisingly straight although the donors used belonged to different substance groups. This straightness indicates that the main factor in charge transfer was actually the ionization potential of the donor. Later Briegleb and Czekalla[54,55] found ways to calculate the charge spectrum in advance and found the same linear relation, their experiments being soon corroborated by Foster.[56] The lines obtained in these plots are so smooth that it is permissible to turn the argument around and read the ionization potential of a donor by plotting the charge transfer spectrum of its complex on the curve of the corresponding acceptor.

If the energy relations greatly favor charge transfer, then the electron could pass spontaneously from D to A. This would correspond to the complexing of a

[52] H. McConnell, J. S. Ham, J. R. Platt, *J. Chem. Phys.* **21**, 66, 1953.

[53] S. H. Hastings, J. L. Franklin, J. C. Schiller, and F. A. Madsen, *J. Am. Chem. Soc.* **75**, 2900, 1953.

[54] G. Briegleb and J. Czekalla, *Z. Electrochem.* **63**, 6, 1959.

[55] J. Czekalla, G. Briegleb, W. Herre, and R. Grier, *Z. Electrochem.* **61**, 537, 1957.

[56] R. Foster, *Nature* **183**, 1253, 1959.

very strong donor with a very strong acceptor, that is, a donor of low IP with an acceptor of high EA. Such cases have been studied by Kainer, Bijl, and Rose-Innes[57-59] as well as Miller and Wynne-Jones.[60] In such a case we can expect the electron to pass spontaneously from D to A, and to resonate between the two molecules, contributing with its resonance energy to the binding forces. The two parted electrons can also impart a dipole moment to the molecule, and, may no more compensate each other's magnetic moment, thus rendering the complex paramagnetic. In the extreme case, the two molecules may even part altogether forming two independent free radicals. Kainer, Bijl, and Rose-Innes have shown that as the difference EA–IP decreases, the molecule complex becomes more and more paramagnetic. Kainer and Überle[61] have shown that such complexes can reach a paramagnetism which corresponds to 40% of the electrons being completely uncoupled. Kainer and Ottig[62] could detect such uncoupling also by IR spectroscopy.

The two cases discussed: IP $\gg$ EA and IP $<$ EA were but extremes. There will be many intermediate cases and as Mulliken has shown, even in one and the same complex, the transfer by photons and spontaneous transfer may be mixed. As the difference IP–EA

[57] H. Kainer, D. Bijl, and A. C. Rose-Innes, *Naturwiss.* **41**, 303, 1951.

[58] H. Kainer, D. Bijl, and A. C. Rose-Innes, *Nature* **178**, 1462, 1956.

[59] D. Bijl, H. Kainer and A. C. Rose-Innes, *J. Chem. Phys.* **30**, 756, 1959.

[60] R. E. Miller and W. F. K. Wynne-Jones, *J. Chem. Soc.* 1959, p. 2378.

[61] H. Kainer and A. Überle, *Chem. Ber.* **88**, 1147, 1955.

[62] H. Kainer and W. Ottig, *Chem. Ber.* **88**, 1921, 1955.

becomes less and less, spontaneous transition will become more and more important, till, eventually, spontaneous transfer may take over altogether and practically a whole electron may be transferred, being lodged more or less permanently on A. The dissociation into two free radicals may make the situation final. Strongly polar solvents will promote the spontaneous transfer and a subsequent dissociation by depolarizing the strong coulombic attractions generated between the two molecules by the transfer of the electron. Weak transfer of the first type will be favored by homoiopolar solvents.[*]

The aforesaid makes it clear also that we have to distinguish between two different events in charge transfer: the bringing of the two molecules together in the desired proximity, and the charge transfer proper. If IP and EA are not favorable, no charge transfer can take place, but there may be cases in which the values of IP and EA are favorable, but no charge transfer can take place because the two molecules do not complex or the desired proximity cannot be achieved owing to steric hindrances. So there may be cases in which there is charge transfer *in vivo*, while there is none *in vitro*. The cell may, for instance, hold two molecules in close proximity, two molecules which would not complex spontaneously. This it could do by binding them to the same enzymic surface or by linking them together by a covalent bond. Nature may also render IP–EA or steric relations favorable by distorting, activating D and A.

To sum up: we could distinguish, schematically, between two types of charge transfer. In the first, the

[*] Refs. 54, 55, page 59.

energy of a photon is needed to transfer the electron, when IP ≫ EA. The sequence of events would be:

$$D + A \leftrightarrow DA, \quad DA + h\nu \rightarrow D^+A^-.$$

In the second type spontaneous charge transfer takes place, we could symbolize events by writing

$$D + A \leftrightarrow DA \leftrightarrow D^+A^- \leftrightarrow D^+ + A^-.$$

As stated, the transition between the two types is gradual. Starting with IP ≫ EA we would need light of high energy to work the transfer. As the difference between IP and EA becomes less, light of increasingly longer wavelength would be needed and spontaneous transfer would become more and more important. So we can expect the plot of frequency versus IP to yield a straight line where much energy is needed for the transfer. However, as we come to longer wavelength, and spontaneous transfer gradually takes over, we can expect the curve to deviate from the straight line and become eventually asymptotic. A slight curvature is noticeable even in Briegleb and Czekalla's curves and becomes evident in the plot of Fujimori[63] (Fig. 11). In this curve, instead of the IP of the various donors their k value was plotted against the frequency of the light absorbed. (The k being a linear function of IP, this should make no difference, except for making the scatter somewhat greater since k is somewhat more uncertain than IP.) The four different curves in Fig. 11 were obtained with four different acceptors. The elevation of these curves corresponds to the difference in the EA of the acceptors used. The curvature becomes the stronger the lower we go and the curve

[63] E. Fujimori, unpublished.

62

suggests that spontaneous transfer becomes an important factor where we approach the infrared, the curve tending to become asymptotic.* Unfortunately, there

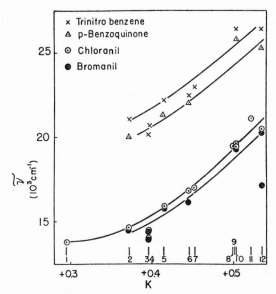

FIG. 11. Frequency of the maximal light absorption as plotted against the k of various donors in complexes formed with four different acceptors. *Key:* 1 = Naphthacene; 2 = Benzpyrene; 3,4 = 9, 10-Methyl-1,2-benzanthracene and 20-methylcholenthrene; 5 = Anthracene; 6 = Pyrene; 7 = Benzanthracene; 8 = Dibenzanthracene; 9 = Picene; 10 = Stilbene; 11 = Chrysene; 12 = Indole.

are too few points in this region because the choice of substances and the technical difficulties, caused by insolubility or color, become rather great. McConnell,

* It might be worth noting that two of the strongest carcinogens, 9,10-dimethylbenzanthracene and 20-methylcholanthrene do not fit into the curves, their points falling somewhat under it. Benzpyrene fits the curve. All of these three carcinogens lie very low.

Ham, and Platt* have already felt the need to extend research in this region.

In the author's opinion, charge transfer may be one of the most important, frequent, and fundamental biological reactions, a possibility, clearly recognized by Mulliken.† It may be objected that there being no light in our body, light-induced charge transfer can have no major biological importance, while substances giving a strong spontaneous charge transfer are rather rare. Most charge transfers which have been studied so far belong to the first type. But, here again, I have to point to the differences between physics and biology. The physicist, when studying charge transfer, will study the interaction of substances which he finds on his shelf, while the biologist works with substances which Nature has developed through millions of years for specific purposes. So the question is not: is charge transfer a common, weak, or strong phenomenon? The question is: does charge transfer belong to Nature's ways, or not? If so, then we can trust that Nature has developed specific substances to fit her purpose. On the subsequent pages, examples of strong charge transfer between biological substances will be given.

The reader may think that the author has dropped into the common error of overemphasizing the importance of a factor which took his fancy. However, he would like to remind the reader of G. N. Lewis' definition of an acid and base. This great pioneer of science defined acids or bases as substances capable of giving up or taking up electrons. So, the acceptor-donor relation of charge transfer merges with the broader acid-

* Ref. 52 on page 59.
† Refs. 45, 46, 47 on page 55 and ref. 22 on page 32.

base concept which is, undoubtedly, one of the foundations of chemistry. The donor-acceptor relation is also related to oxidoreduction, which is one of the cornerstones of biology. So charge transfer makes part of a most basic triad.

VI

Problems of Charge Transfer

WHAT DO WE GAIN BY INTRODUCING CHARGE TRANS-
fer into the complex texture of biochemistry and biophysics?

Charge transfer allows us to transfer an electron
from one substance to another without major loss in
energy, since it does not involve a rearrangement in
molecular structure.

It brings into the realm of biological oxidation the
great host of substances which are capable of giving
off but one electron, which do not affect an electrode
and were hitherto not regarded as redox agents at all.

Charge transfer brings into play the excited levels
which may not have been available before because the
energies needed to raise an electron into the excited
level of the same molecule are, as a rule too great.

By charge transfer relatively inactive molecules may
acquire a high reactivity. The donor, having developed a "hole" in its low-lying ground level, becomes a
good acceptor, while the acceptor having acquired an

67

electron on its high-lying excited orbital, becomes a good donor. A charge transfer complex is something between a regular closed shell molecule and a free radical, and the great reactivity of free radicals need not be emphasized. If the complex dissociates, as happens in extreme cases, two real free radicals are formed.

An acceptor may accept an electron also from a saturated energy band, thus creating a hole and making the band conductant. Conversely, a donor may donate an electron to an empty energy band, rendering this band conductant. So charge transfer opens the way for semiconduction into biology. As shown by Tollin,* sheets of donor molecules layered over sheets of acceptor molecules become conductant and show a strong photoelectric effect.

Akamatu *et al.*[63] found that perylene, violanthrene, and other higher aromatics undergo marked increases in electrical conductivity when complexed with I_2 or Br_2. These complexes show a marked paramagnetic component (Matsunaga[64]). This work demonstrates that an insulator or poor semiconductor can be transformed into a good one and that I_2 may also give strong charge transfer complexes with the spontaneous passage of almost a whole electron.

How can we distinguish between "strong," spontaneous, and "weak" light-induced charge transfer?

* Ref. 39 on page 52.

[63] H. Akamatu, H. Ivokuchi, and Y. Matsunaga, *Bull. Chem. Soc. Japan* **29,** 213, 1956; *Nature,* **173,** 168, 1954.
[64] Y. Matsunaga, *Bull. Chem. Soc. Japan* **28,** 475, 1955; *J. Chem. Phys.* **30,** 855, 1959.

Having no light in our body this distinction may be important. How can we prove charge transfer at all? There are various methods.

OPTICAL METHODS. Most charge transfer complexes are strongly colored and their spectrum can give information about the nature of the underlying reaction. If the absorption shows the relation between frequency and ionization potential, described in the previous chapter, then, evidently, we are dealing with "weak charge transfer," induced by the light absorbed. If, on the contrary, there is no such relation and the spectrum shows some relation to the (perhaps somewhat shifted or distorted) spectrum of the free radical of one of the complexing substances, then, evidently, we have strong charge transfer in hand with a spontaneous transition of electrons and a trend of going into the ionic state or into free radicals. Many complexes will show a mixed behavior.

MAGNETIC METHODS. In valency saturated compounds electrons occupy orbitals in pairs. Electrons have their spin, and a spinning electron is a tiny magnet. Since the two electrons occupying the same orbital always spin in opposite directions, they cancel out each other's magnetic moments. However, if the two electrons become separated, as may be the case in a strong charge transfer, they may no longer compensate one another, rendering the substance paramagnetic which can be expected to declare itself in magnetic measurements. There is also another possibility which may render the complex paramagnetic: as soon as the two electrons are no longer strongly coupled, and occupy different orbitals, they become relieved of the limitation of the Pauli principle according to which two electrons, forming a pair on the same orbital must spin in opposite directions. If one of the two uncoupled electrons reverts its spin, the complex goes into the triplet state, adding a paramagnetic component. McGlynn and Boggus[65,66] actually found in photolytic experiments such complexes to give a triplet emission.

There are two methods for detecting paramagnetic behavior: the magnetic balance and the electron spin resonance (ESR).[67-70] The former is rather crude as compared to the latter. The signal given in ESR depends on circumstances. Free radicals give a sharp and high ESR signal extending over 10–50 gauss. If the electron is close to other unpaired electrons, it will be perturbed and may give a broad signal. Exposed to the magnetic influence of different nuclei it may split up giving a "hyperfine" structure. Although the ESR method is rich in pitfalls, it is one of the most powerful tools of research finding access, now, to biological laboratories. How far a charge transfer complex will give an ESR signal depends on circumstances. In a weak transfer the two electrons remain strongly coupled and so give no ESR signal, or, give a signal only under strong illumination.

The situation will be different in strong charge transfer, depending on how strongly the transferred elec-

[65] S. P. McGlynn and J. D. Boggus, *J. Am. Chem. Soc.* **80,** 509, 1959.
[66] S. P. McGlynn, *Chem. Rev.* **58,** 1113, 1958.
[67] For technique and theory see: G. E. Pake, S. I. Weissmann, and J. Townsend, *Disc. Faraday Soc.* **19,** 147, 1955.
[68] W. Gordy, W. V. Smith, and R. F. Trambarülo, "Microwave Spectroscopy," Wiley, New York, 1953.
[69] P. B. Sogo and B. M. Tolbert, *Biol. and Med. Phys.* **5,** 1, 1957.
[70] D. J. E. Ingram, "Free Radicals," Academic Press, New York, 1958.

tron is still coupled to its partner, on how "strong" this charge transfer is (that is, how big a portion of the electron is transferred), what proportion of its time it spends on the acceptor,* and how far the complex tends to split up, going into the ionic state in which the dissociated partners are free radicals.

The ESR is a wonderful instrument which gives unique information. Its handling requires specific knowledge and experience, complicated (and expensive) machinery, and a great deal of caution in interpretations. One can make a discovery with it in one day, but, then, it takes six months to disprove it. Watery solutions, which might interest the biochemist most, are difficult to study, because only tiny amounts of this solvent will not blur the spectrum. Impurities, present in traces, may mislead the researcher. Special caution is required when dealing with solid state. Distortions of lattices or their physical damage may lead to signals or free radicals. The theory of ESR is still in its infancy; most of the substances which have been studied are simple in nature and surprises may be expected. There appears to be no theory for the explanation for the broad signals reported by Blumenfeld, *et al.*[71,72] in nucleoproteins.†

---

* Such expressions as "transfer of part of the electron" or "the electron spending part of its time on the new orbital" must be taken with a grain of salt. They do not represent reality, only the shadow of it. These partial reactions, in any case, try to describe a state between the original and the final one in which an electron may be definitely lodged on its new orbital.

† See footnote (†), page 72.

[71] L. A. Blumenfeld, A. E. Kalmanson, and Shen-Pei, *Gen. Doklady Akad. Nauk, USSR,* **124,** 1144, 1959.
[72] L. A. Blumenfeld, *Biophysika* **4,** 515, 1959.

Kainer, Bijl, and Rose-Innes,‡ who were the first to apply ESR to charge transfer complexes, have shown that when going from weak charge transfer to an increasingly stronger one, by taking as donor stronger reductants, and as acceptor stronger oxidants, the ESR signal appeared and became gradually stronger.

DIPOLE MOMENT measurements can also support charge transfer, and have been measured by Weiss,* Briegleb, and Czekalla,** Kainer and Überle,# and others. Unfortunately, the measurement of dipole moments is not simple, especially not in polar solvents which favor strong charge transfer (while weak charge transfer is more favored by homoiopolar solvents). Also, the conclusions deduced from dipole moments may be less straightforward than those deduced from optical and magnetic measurements. Crystallography of charge transfer complexes may also give additional evidence. (Planar aromatic molecules can be expected to lie parallel with their centers shifted to allow charge transfer otherwise excluded for symmetry reasons.§)

Does charge transfer open new possibilities for the understanding of oxidation or other biological phenomena?

† It may be added that I. Isenberg and the author found in preliminary studies most tested steroids (Δ^4-androstene-3, 17-dione; Δ^5-androstene-3, 17-dione; Δ^5-androstene 3β, 17β-diol, cholesterol; progesterone; stigmasterol, and deoxycorticosterone) to give a broad and low signal while testosterone was negative. Further studies are needed to decide whether this signal was due to impurities, or other disturbing factors, or was a property of these steroids perhaps complexed with O_2.

‡ Refs. 57, 58 on page 60.
* Ref. 42, page 54.
** Ref. 54, page 59.
# Ref. 61, page 61.
§ Ref. 46, page 55.

Let us suppose that D, A_1, A_2, and A_3 in Fig. 12 represent four molecules in close proximity, and suppose that D transfers an electron from its ground level to the excited level of A_1. This electron could not move on being held in place by the electrostatic attraction between D and A_1. This situation would change if, from an outside molecule, we would transfer one elec-

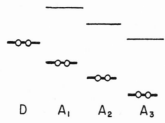

FIG. 12. Symbol of a possible quantum-mechanical framework of electron transport. Thick lines: ground states; thin lines: first excited levels. See text.

tron to D, filling the "hole" in its ground level. The electron transferred earlier to A_1 would become released from its electrostatic bondage and could cascade down over the excited levels of A_2 and A_3. It is not impossible that something like this really happens in biological oxidation. It would be equally possible to remove one electron from the ground state of A_3, whereupon an electron of A_2 could fill the hole, while A_2 would take an electron from A_1, from D, all the electrons cascading one step down.

We could also suppose, as a fascinating possibility, that we transferred an electron from the ground state of D to the excited level of A_3 and then transfer an electron from the ground state of A_3 to an outside molecule, say O_2. In this case the electron could fall down from the excited level of A_3 into its ground level, filling the empty hole on it, and emitting its excess energy in the form of a photon. Some such possibility may underlie bioluminescence. All this is very amusing because it makes a biochemistry without chemistry. There being no rearrangements in molecular structure or bonds, there are no "chemical changes" and the molecules involved serve merely as a quantum-mechanical framework on which the electrons can move. Maybe the cycle of biological oxidation represents some such scaffolding.

Summing up we can thus state that charge transfer opens diverse new, important, and intriguing possibilities.

VII

Three Examples of Charge Transfer

AS MENTIONED BEFORE, IN THE FORMATION OF A charge transfer complex we must distinguish between two events: bringing the molecules together and transferring charge. So, if we want to study charge transfer proper, we must find methods for holding the two molecules close to one another.

There are various possibilities. We may build the two substances into a crystal, in which case the lattice forces will hold the two molecules together. We may simply evaporate the solvent in which we dissolved both substances in question. The evaporating solvent will leave the two substances behind in close proximity. Another method consists of freezing the watery solution of the mixture. The water will, in this case, crystallize out and leave the dissolved molecules in intimate touch. In this case complex formation will be favored also by the decreased temperature. In the living cell, also, association and charge transfer may be promoted by the trend of the matrix, water, to form structures which limit the degrees of its freedom (see next chapter). On the subsequent pages I will give examples of charge transfer produced by these methods.

QUINONE-HYDROQUINONE

A classic material for the study of weak charge transfer is quinhydrone, formed by mixing quinone and hydroquinone. Both of these molecules are flat and have an extensive system of π electrons. So, if they come to lie close enough to one another with their faces parallel, then the orbitals of their π electrons may overlap, forming what Dewar[73] calls a "π-π-bond" in a "π complex." Osaki and Matsuda* studied the crystal structure of quinhydrone by X-rays and found that the two rings, that of the hydroquinone and quinone, lie very close to one another. While the distance usually found between associating aromatic compounds is 3.5–5.7 Å, in quinhydrone it was found to be 3.16 Å, a very close proximity, indeed. In the crystal, the quinone and hydroquinone molecules lie alternating, on top of one another, with their centers slightly shifted, their plane being inclined by 34° toward the long axis of the needle-shaped crystals. So the electrons, excited by light, may oscillate parallel to the plane of the molecules or at right angles thereto, oscillating between the neighboring quinone and hydroquinone molecules which thus form a charge transfer complex. According to the light vector, whether more parallel or at right angles to the plane of the molecules, the quinhydrone will thus absorb light of different wavelength, will be dichroic, as shown by Nakamoto.[74] This can easily be demonstrated by drying down an acetone solution of

* Quoted from Nakamoto, ref. 74 on page 76.

[73] M. J. S. Dewar, *Nature* **156,** 784, 1945. "The Electronic Theory of Organic Chemistry," p. 62, Clarendon Press, London, 1949.

[74] K. Nakamoto, *J. Am. Chem. Soc.* **74,** 1739, 1952.

quinhydrone on an object slide. Under the microscope (Fig. 13) one finds featherlike crystals running in different directions. If these are observed in polarized light (Fig. 14) part of the crystals will be found to be dark blue, while others, running at right angles to the former are light yellow. If the preparation is turned by 90° the colors interchange. (In Fig. 14 the blue ones appear black.) The absorption of quinhydrone shows two maxima,[75] one at 550 and one at 380, corresponding to the two different oscillations.

A 10^{-2} M p-hydroquinone solution is colorless and a 10^{-2} M p-quinone solution is light yellow. On mixing the two, the color hardly changes, though 10^{-2} M quinhydrone can be expected to be red-brown. Our mixture remains colorless (first tube on the left in Fig. 15) because, at this dilution, quinhydrone dissociates, or rather, is not formed at all.

Amusing experiments can be made with this 10^{-2} M mixture. If one suddenly freezes this solution,* it turns into dark ultramarine blue (last tube on the right in Fig. 15). This shows that we have, by freezing, forced the molecules to form a complex, and forced them into an unusual relation which favors the oscillation between quinone and hydroquinone molecules at right angles to their plane.† In order to trap the quinhydrone in its red-brown modification, we have to add to the

---

* As a freezing mixture I usually use powdered dry ice suspended in the monomethyl ester of ethylene glycol.

† If somewhat more concentrated solutions are frozen, then on melting, the ice leaves behind dark blue, almost black, fine needle-shaped crystals which show no dichroism, contrary to the regular quinhydrone crystallized at room temperature, which is red-brown and dichroic.

[75] Keisuke Suzuki, *Busseiron Kenkyu*. Research on the Structure of Matter, No. 102, page 5, Japan.

Fig. 14. Quinhydrone in polarized light.

Fig. 13. Quinhydrone crystals in plain light.

Fig. 15. From left to right: First—mixture of 0.01 M p-quinone and hydroquinone; second—mixture of 0.01 M quinone, hydroquinone, and 10% glucose, frozen; third—same after slight warming; fourth—0.01 M quinone and 0.01 M hydroquinone, frozen. (Details—see text.)

mixture of one part of $10^{-2} M$ quinone and quinhydrone, one part of 10% glucose. By sudden freezing we can then trap the quinhydrone molecules in their red-brown form which is metastable in our frozen system (second tube from left). That this is so, can be demonstrated by dipping the frozen tube, for a few seconds, into water of room temperature. The carrot-red color shoots over into ultramarine blue (third tube in Fig. 15). One can amuse oneself (and one's children) with these colorful experiments. All these changes are reversible and the highly colored ice melts into a practically colorless liquid.

What holds the quinhydrone molecule so tightly together in the quinhydrone crystal is, in the first place, not the H-bonds but the π-π interaction.[76] In the blue quinhydrone this interaction seems to be favored. Simply drying down a mixed solution of quinone and hydroquinone will result only in the formation of brown quinhydrone, as can easily be demonstrated by

[76] That the essential point in quinhydrone interaction is not the H-bonding has been shown by Michaelis and Granick (*J. Am. Chem. Soc.* **66**, 1023, 1944), who demonstrated the formation of quinhydrone also when the H's in hydroquinone were exchanged for methyl groups. That there is no obligatory exchange of H's between the two molecules in the crystal has been demonstrated by I. P. Gragerov and G. P. Mikluhin (*Doklady Akad. Nauk. USSR* **62**, 79, 1948) and A. Bothner-By (*J. Am. Chem. Soc.* **73**, 4228, 1951) who showed that if isotopically marked quinone formed quinhydrone with unmarked hydroquinone (or vice versa), then from the crystals the marked or unmarked components could be recovered without having taken over H from, or given H's to, their partner. Quinhydrone formation was thus, in these cases, purely an interaction of the π electron systems. In these experiments the H-atoms of the benzene ring were substituted to prevent a rearrangement of double bonds.

pouring a few drops of a methanoli solution of quinhydrone on filter paper: the evaporating solution leaves a brown patch behind. If this patch is wetted with water and frozen by placing some powdered dry ice on top, the color changes into a vivid blue.

Quinhydrone is a classic example of a weak charge transfer. Quinone is a good acceptor but hydroquinone is a poor donor (see Table I) and so there is no spontaneous transfer of electrons. The orbitals of the very closely packed quinone and hydroquinone molecules overlap and the light makes the electrons oscillate between the neighboring molecules if the light vector moves them in this direction. Neither the blue, nor the red-brown quinhydrone gives an ESR signal. As mentioned before, charge transfer is very sensitive to distance. The flat benzene rings fit very closely and if no bulky side chains are introduced there is no steric hindrance to disturb close proximity.

RIBOFLAVIN (FMN) AND SEROTONIN

As is generally known, FMN is intensely yellow and its solutions show a strong green fluorescence which, on freezing, changes into an orange phosphorescence.[77] Serotonin (5-hydroxytryptamine) is a close relative of tryptophan. It has no color. It is not a "reducing agent" having no two stable states which would correspond to the loss of two H's or two electrons. However, it has a fairly high filled orbital ($k = 0.486$) and can be expected to be a fair monovalent electron donor while FMN with its lowest empty orbital ($k = 0.343$) is a good acceptor. So, if the two molecules are kept close together a charge transfer could occur, an electron

---

[77] A. Szent-Györgyi, "Bioenergetics," pages 28, 29. Academic Press, New York, 1957.

Fig. 16. From left to right: First—$5.10^{-4} M$ riboflavin phosphate Na (FMN); second—mixture of $10^{-4} M$ FMN and serotonin; third—$5.10^{-4} M$ FMN, frozen; fourth—FMN—serotonin mixture, frozen; fifth—same at acid reaction. (Details— see text.)

passing from the indole to the vitamin. Figure 16 shows the color changes taking place on freezing of an aqueous solution containing a mixture of the two substances. The first tube on the left contained a 5.10^{-4} M solution of FMN, while the next tube contained, in addition, an equivalent amount of serotonin. There is no striking difference in color between the two although the mixture may be a trace darker. The third tube contained pure FMN frozen to show that freezing, as such, makes no difference in color: the ice is intense yellow. The fourth tube shows the FMN-serotonin mixture frozen. Its color is mahogany brown.[78] The last tube contains the same[79] mixture frozen, containing also 1% strong HCl. It is black. Most striking changes indeed. All of them are reversible: on melting, the original color returns. Freezing promotes a reaction occurring to a smaller extent also at room temperature, as can be shown by the noticeable darkening of the FMN solution obtained on mixing FMN and serotonin of higher concentration.

With the method of Strehler and McGee[80] the absorption spectrum of the frozen solution can be measured. The brown solution shows an absorption maximum around 500 mμ. What makes this mahogany color and absorption interesting is its similarity to the color of the free radical of FMN. This radical, Michaelis' rhubroflavin, in free condition, is stable only at a strongly acid reaction and can readily be produced by reducing FMN by dithionite in presence of strong HCl. This free radical, produced by the monovalent

[78] I. Isenberg and A. Szent-Györgyi, *Proc. Natl. Acad. Sci. U.S.* **44**, 857, 1958.
[79] I. Isenberg and A. Szent-Györgyi, *Ibid.* **45**, 1229, 1959.
[80] B. Strehler and M. McGee (unpublished), quoted from ref. 79.

reduction of FMN, was discovered a quarter of a century ago by Kuhn and Wagner-Jauregg,[81] and has been studied since, extensively, by Michaelis and his associates,[82-84] and Beinert.[85,86] It has its absorption maximum, depending on pH, between 475 and 570 mμ. In partially reduced yellow enzymes Beinert found a strong absorption at 500 mμ. This means that we can actually identify, with great probability, the molecular species which is responsible for the absorption in our neutral frozen mixtures as being related to the free radical of FMN.

An absorption, similar to that of the free radical, can be observed even at room temperature in stronger solutions of pure FMN without an added donor. FMN being both a good acceptor and a fair donor (k of the highest filled orbital $= 0.496$) we can expect a charge transfer from one molecule to the other in the FMN dimers.* As pointed out by Mulliken, transfer is possible not only between different, but also between identical, molecules and Michaelis and Granick[87],† as well as Beinert,[85] demonstrated the dimerization.

E. Haas,[88] Beinert,[85] and Ehrenberg and Ludwig[89]

* The donor properties of FMN can also be demonstrated by the color change produced on freezing of FMN solutions containing some I$_2$. Molecular iodine with its unsatisfied electronegativity is a good acceptor. On freezing, the solution turns black.

† Refs. 48, 49, 50, on page 55.

[81] R. Kuhn and T. Wagner-Jauregg, *Ber.* **67**, 361, 1954.
[82] L. Michaelis, M. P. Schubert, and C. V. Smythe, *J. Biol.* **116**, 15, 1936.
[83] L. Michaelis and G. Schwarzenbach, *J. Biol. Chem.* **123**, 521, 1938.
[84] L. Michaelis, Theory of Oxidation and Reduction, *in* "The Enzymes," Vol. **II**, Part I, J. B. Sumner and K. Myrbäck (Eds.), Academic Press, New York, 1951.
[85] H. Beinert, *Biochim. et Biophys. Acta.* **20**, 588, 1956.
[86] H. Beinert, *J. Biol. Chem.* **225**, 465, 1957.
[87] L. Michaelis and S. Granick, *J. Am. Chem. Soc.* **66**, 1020, 1944.
[88] E. Haas, *Biochem. Z.* **290**, 291, 1937.
[89] A. Ehrenberg and J. D. Ludwig, *Science* **127**, 1177, 1958.

demonstrated the formation of such red radicals also in enzyme-linked FMN and FAD, under influence of metabolic H-donors or reducing agents, while Commoner[*] and his associates, as well as Ehrenberg and Ludwig,[†] showed FMN to give ESR signal on partial reduction, and Ball[90] noted the brown color on xanthin oxidase, a flavoprotein.

The black FMN-serotonin mixture, frozen at acid reaction, shows in the spectroscope[‡] a broad absorption stretching over the whole visible and having a maximum in the near ultraviolet, with its tail in the visible, similar to the absorption of the green FMN radical, Michaelis' "verdoflavin." Another maximum is located at 570 mμ with a shoulder at 620 mμ. Evidently, various spectra become confluent here, maybe that of FMN$^-$, and FMN$^-$ with one or two protons bound (as suggested by calculations of Karreman[§]), possibly with some charge transfer spectrum mixed in. In any case, there can be little doubt that this is a very strong charge transfer with nearly a whole electron transferred. In frozen mixtures containing equivalent amounts of FMN and serotonin, the lack of any light emission indicates that the whole FMN has been involved in the formation of the charge transfer complex, at both neutral and acid reactions, one whole electron having been transferred to it.

The transferred electron may have come from the π

[*] Refs. 7, 8 on page 6.
[†] Ref. 89 on page 82.
[‡] Ref. 79 on page 82.
[§] G. Karreman, Personal communication.

[90] E. Ball, *Cold Spring Harbor Symp. Quant. Biol.* **7**, 100, 1939.

pool of the serotonin or from the lone electron pair of its N, which also makes part of the π pool of the molecule.[91]

The assumption that the color changes are due to the absorption of the FMN^- formed and not to a charge transfer spectrum, can be supported by freezing FMN in presence of KI. The iodide anion is a fair electron donor. Although its ionization potential is very different from that of indoles,[*] the color changes obtained on freezing are the same as those produced by serotonin, since the absorbing molecular species, FMN^- are the same. If the spectra had been charge transfer spectra and not the spectra of the radical one would have expected marked changes in absorption when using different donors. With the FMN complexes the spectra were practically unchanged.

There is one point about the interaction of FMN and serotonin which demands attention: its great intensity. Although serotonin is a somewhat stronger donor than indole, or some of its derivatives such as tryptamine, this extraordinary reactivity is characteristic for the indole formation in general and is in no way explained by the energy values of the highest filled molecular orbital. According to its k (0.534) indole is only a "fair" donor, not a really good one. Its charge transfer spectra correspond to this k, as shown by the curves of Fujimori (Fig. 11). However, this curve

[*] The electron affinity of an atom is equal to the ionization potential of its monovalent ion. The electron affinity of I and thus the ionization potential of I$^-$ is 3, 14 EV. That of indoles is probably considerably higher.

[91] As shown by Fujimori, part of the emission of tryptophan disappears in presence of strong acid indicating that the excited electron came from a lone electron pair (*Biochim. et Biophys. Acta* **40**, 257, 1960).

shows that with bromanil, indole gives a second absorption peak which corresponds to a considerably lower frequency. Such a double peak, according to Kainer, Bijl, and Rose-Innes,[*] is characteristic for very strong charge transfer, the interaction of strong oxidizing and reducing agents. But even the k, corresponding to this second absorption peak[†] does not explain the observed strong donor property. There seem to be other hitherto unknown molecular indices which entail this extraordinary reactivity. There is but one general statement we can safely make about all this, and this is that if Nature developed a substance for electron transfer, then she may have given to this substance extraordinary qualities as an electron transmitter. Indole has such qualities which make it likely that Nature developed this ring actually for services of this kind. This seems to be noteworthy because physiologically rather active and important substances are found among indole derivatives. The plant hormone indole acetic acid, and serotonin, may be quoted, but the most remarkable fact to me is that protein also has an indole side chain contained in its tryptophan. This may throw light, some day, on the real meaning of protein. It is tempting to think that Nature introduced this amino acid into the protein molecule to mediate its charge transfer reactions.[‡][92]

[*] Refs. 57, 58 on page 60.

[†] This absorption would correspond to a k of 0.46.

[‡] Myosin and actin, which transform chemical energy into motion, are rich in tryptophan while paramyosin, which has a passive role, supplying a catch mechanism with its crystallization, has none. The highly crystalline tropomyosin, which seems to have some similar function, is also devoid of tryptophan.

[92] W. H. Johnson and Andrew G. Szent-Györgyi, *Science* **130**, 160, 1959.

FMN too has but a fairly low k for its lowest empty molecular orbital (-0.344) and its reactions as an electron acceptor are stronger than this k allows us to expect. So what has been said for indole holds, also, for FMN of which we do know that Nature made it to serve as an electron transmitter.[*]

According to its k, FMN is not an especially good donor (k of the highest filled orbital $= 0.500$), but it should be remembered that the H-donor in oxidation is not FMN, but FMN$^-$, which has an electron accepted on an antibonding orbital and can be expected to be an especially good donor. This is the case as has been shown for FMNH$_2$ by the Pullmans,[†] the k value for the highest filled orbital having a negative value, which was the first instance observed of a filled orbital with an antibonding character (see Table I).

Other donor-acceptor pairs of biological substances have been described by Cilento and Giusti,[93] and Isenberg and the author.[‡] The electron transmission between serotonin and DPN$^+$ can be beautifully demonstrated by the freezing technique: on freezing, the mixture assumes the yellow color of DPNH. Fujimori[94] studied charge transfer, using pteridines as an acceptor.[95,96] The first to point out the possible role of

---

[*] Karreman (personal communication) showed that the FMN and indole molecule can be laid on top of one another in a way that positively charged C atoms come to lie opposite negatively charged ones, which may contribute to attractions between the two molecules and favor charge transfer. The strong interactions observed, however, cannot be ascribed solely to this factor since indole acts as a strong donor, also, with other acceptors.

[†] Ref. 25 on page 44.

[‡] Ref. 79 on page 81.

[93] G. Cilento and P. Giusti, *J. Am. Chem. Soc.* **81**, 3901, 1959.

[94] E. Fujimori, *Proc. Natl. Acad. Sci. U.S.* **45**, 133, 1959.

[95, 96] H. A. Harbury and K. A. Foley, *Proc. Natl. Acad. Sci. U.S.* **44**, 662, 1958; *Ibid.* **45**, 178, 1959. These authors measured the dissociation constants of complexes of isoallaxizins in various solvents and leave it open as to how far charge transfer was involved.

charge transfer in the function of DPN, and hence in biological oxidation, was Kosower.[*]

CORTISONE I_2

The interaction of hydroquinone and quinone was a classic $\pi-\pi$ interaction. The interaction of serotonin and FMN also involved, in all probability, the π pool of these substances, though the transferred electron might have been derived from the lone pair of N. In these cases the planarity of the interacting molecules greatly favored complex formation, the two flat surfaces providing approachability and a free play of the attractive forces leading to intimate union.

Mulliken[†] has pointed out that charge transfer complexes can be formed not only by an extensive planar system of conjugated double bonds with their π pools, but also by local donors or acceptors, atoms or small atomic groups making part of bigger molecules. The lone pair of N or O molecules can serve as local "onium donors." Miller and Wynne-Jones[‡] have shown the NH_2 groups to be strong donors.

Reid and Mulliken[97] studied such a case: the charge transfer between pyridine and I_2. As they point out, the various solvents dissolve I_2 with two different colors. The homoiopolar ones, such as chloroform, dissolve it with a violet color, which is also the color of the iodine vapor. The heteropolar solvents, as alcohols, dissolve it with a brown color (which in higher dilution looks rather yellow). This color change they ascribe to a charge transfer which shifts the absorption

[*] Ref. 15, page 19.
[†] Ref. 47, page 55.
[‡] Ref. 60, page 60.

[97] C. Reid and R. S. Mulliken, *J. Am. Chem. Soc.* **76,** 3869, 1954.

of I_2 from the violet 520 mμ to 410–370. Pyridine also dissolves I_2 with a yellow color, and at the same time a "charge transfer spectrum" appears in the UV. The donor, in this case, is the lone pair of the N atom of pyridine. According to Mulliken and Reid, first an "outer complex" is formed which then goes over into the "inner one"; in the cortisone complex, also, some such rearrangement may occur.

As mentioned earlier, charge transfer is very sensitive to distance and a very close proximity of acceptor and donor are a *sine qua non*. It is easy to see that two flat aromatic structures can easily establish such proximity with their faces parallel, and conditions will favor charge transfer if the donor is the π electron system of the one while the acceptor is the π system of the other. The situation will be different if we are dealing with a localized donor, as was the case with pyridine in which the N atom donated one of the electrons of its lone pair. In this case the desired proximity can be established much easier if the acceptor itself is but a small molecule such as that of I_2 which can easily find the desired steric position in relation to the N atom. Charge transfer will be more difficult to achieve between such a local donor and the π system of a complex molecule.

Ketosteroids are big flat molecules with CO groups in which the O, with its lone pair of electrons, may act as a strongly localized donor. So in order to find out whether it really will do so, we do well to choose a small molecule, such as I_2, as acceptor. The question is only how to bring the O and I_2 together. Both I_2 and steroids are soluble in chloroform, but nonpolar solvents disfavor strong charge transfer. Strong charge transfer is favored by polar solvents. However, most polar solvents, like alcohol, which dissolve both steroids and I_2, are fair local electron donors themselves and so can be expected to compete with the steroid for I_2, with which they form charge transfer complexes themselves. Water is a strongly electropolar solvent and is a very poor donor, so, in principle it would suit our purpose, but it dissolves neither steroids nor I_2 very well. So, evidently, some trickery is needed. We can try, for instance, to dissolve both steroids and I_2 in chloroform, pour some of the solution on filter paper and let the chloroform evaporate, expecting the I_2 and steroid to form a complex. Then we can wet the paper, thus exposing the complex to the desired heteropolar solvent, expecting a charge transfer to take place and declare itself by a change in spectral properties.

If 0.01 M I_2 is dissolved in chloroform and 0.01 M cortisone is added, there is no change in color. If the solution is poured on (Watman 1) filter paper and the chloroform is allowed to evaporate, a yellow complex is left behind (without cortisone the I_2 evaporates too, leaving a colorless filter paper behind).

If the paper is dipped, now, into water, its yellow color turns deep blue. The color will be even more intense if instead of water the paper is dipped into a Lugol solution,* diluted with 25 volumes of water.†

* Lugol's solution contains 5 grams iodine and 10 grams KI in 100 ml water. A Lugol diluted 1/25 colors the paper itself but faintly brown.

† Acridine, which is not only a good acceptor ($k = -0.343$) but also a good donor ($k = +0.494$) gives a similar color under similar conditions.

The spectroscope shows a broad absorption with a maximum at 740 mμ, where neither I_2 nor cortisone absorbed. The absorption is structureless, has thus all the earmarks of a charge transfer spectrum, lying also at a longer wavelength than the absorption of the reactants. Kendall's compound S shows a similar, though weaker color change. The color developed by the cortisone-iodine compound is rather unstable and disappears in a few minutes. (Mulliken noted that many charge transfer complexes have a strong tendency to give secondary reactions, as going from "outer" to "inner" complexes, or giving irreversible changes.) Ten per cent methanol completely inhibits the development of the blue color (the alcohol being a not too poor electron donor).

The blue color, with the long wavelength absorption on the border of the infrared, indicates that relatively little energy was needed to transfer an electron from cortisone to I_2. So, with a stronger electron acceptor a charge transfer in the ground state could be expected.

The other ketosteroids studied, give, under similar conditions, an orange color which is more stable than the blue one given by cortisone. Here too the peaks are broad and structureless. The maximum lies with testosterone at 410 mμ, with deoxycorticosterone at 430 mμ, with Δ^5-androstene-3,17-dione at 420 mμ, with Δ^4-androstene-3,17-dione at 450 mμ, and with progesterone at 405 mμ. Cholesterol, stigmasterol, Δ^5-androstene-3β,17β-diol, or estradiol, having no CO, do not show such color changes. Estradiol having a system of conjugated double bonds with their π pool, gives a charge transfer with trinitrobenzene, as shown by Williams-Ashman, while the guest of my laboratory.

The I_2 compounds can be produced preparatively.*

The color changes, given by steroids with I_2, have been described a decade ago by Zaffaroni[98,99] and his associates, who used them for the detection of steroids in chromatography. At that time these reactions were demonstrated to me by Dr. Oscar Hechter of the Worcester Institute for Experimental Biology and Medicine. Charge transfer being yet unknown to me, I could make no sense of these color changes. It was, evidently, the memory of this early demonstration which led me back to these reactions.†

Although there is no final proof for it, the probabilities are that these reactions of steroids and I_2 are charge transfer reactions indicating a strong donor ability of ketosteroids. The experiments of Talalay, Williams-Ashman, and Hurlock, to which I will come back later (Chapter XII) actually show that steroids can act as electron donors and acceptors, mediating the electron transfer between pyridine nucleotides in the presence of a catalyst.

* Dissolve 0.5% cortisone in hot water, then add 1/20 vol. of Lugol's solution. On cooling, the deep blue iodine complex of the steroid separates, can be spun off, washed with little water and pressed out between filter papers.

† For the generous supply of DOC and cortisone, I am grateful to Merck; for the progesterone and compound S and estradiol, to Dr. O. Hechter and the Worcester Foundation for Experimental Biology and Medicine; for the androstenes and other highly purified sterols, to Prof. Charles B. Huggins.

98 A. Zaffaroni, R. B. Barton, and E. H. Keitman, *Science* **111,** 6, 1950.

99 R. B. Barton, A. Zaffaroni, and E. H. Keitman, *J. Biol. Chem.* **188,** 763, 1951.

VIII

Miscellaneous Remarks

WATER

IN MY LITTLE BOOK "BIOENERGETICS" TWO CHAP-
ters were devoted to water. In the present book hardly
any mention has been made of it. This is by no means
because the author has changed his mind about structured water which, so to say, is half of the living
machinery, and not merely a medium or space-filler.
He still thinks that it is a mistake to talk about proteins, nucleic acids or nucleoproteins *and* water, as
if they were two different systems. They form one
single system which cannot be separated into its constituents without destroying their essence. I am more
convinced than ever that half of the contractile matter of muscle is water, contraction the collapse of its
structure, induced by actomyosin. Biology has forgot-
ten water as a deep sea fish may forget about it. If
water was hitherto passed by in silence in the present
booklet this was because the author had no new experimental material to present, although there has
been some progress, and also criticism in different
quarters.

In my first little booklet I talked about "ice" formed around various molecules. J. D. Bernal* justly objected to the term "ice" which means something very definite for a crystallographer. Instead of ice one should rather speak about structured water, or water with restricted freedoms. All the same, "ice" is a nice and short word. It was probably for this reason that it has been used by outstanding earlier investigators like Frank and Evans.[100]

Speaking about progress made in the study of water in its relation to biological structures, the work of I. Klotz[101-103] should not remain unmentioned. Klotz showed that certain reactions of proteins could be explained by supposing the existence of a strongly bound sheet of water around the particles which greatly modified their accessibility,[101] reactivity,[102] and even mediate energy transmission.[103] Eigen's work[104] on the high proton conductivity of ice could bear on such a possibility.[105] Bernal's[106,107] latest work shows that even random liquids have their geometry.

Water may be important for biological happening both by its presence and its absence. Green† stresses

* Personal communication.
† Ref. 21 on page 30; ref. 28 on page 48.

[100] H. S. Frank and M. W. Evans, *J. Chem. Phys.* **13**, 507, 1945.
[101] I. M. Klotz, *Science* **128**, 815, 1958.
[102] I. M. Klotz and R. G. Heiney, *Proc. Natl. Acad. Sci. U.S.* **43**, 74, 1957.
[103] I. M. Klotz, *J. Am. Chem. Soc.* **80**, 2123, 1958.
[104] M. Eigen and L. de Maeyer, *Proc. Roy. Soc.*, A. **247**, 505, 1958.
[105] Review on the possible biological role of water structures: See P. L. Privalov, *Biofizika*, **3**, 738, 1958. Translation: *In* Biophysics **3**, 691, 1958. "The state and role of water in biological systems."
[106] J. D. Bernal, *Nature*, **183**, 141, 1959.
[107] J. D. Bernal, *Nature*, **185**, 68, 1960.

that the process of oxidative phosphorylation in mito-chondria takes place in an anhydrous, lipid matrix. This anhydrous nature of the medium may alter happenings and energy relations. It has been shown by B. Pull-man,[†] for instance, that the hydrolytic free energy of ATP depends, to a great extent, on the dielectric prop-erties of the medium, and, as pointed out by Isenberg,[‡] can be expected to be considerably higher in an an-hydrous than in an aqueous milieu. There may be proc-esses, such as electron transfer, which simply might not work in a watery medium, as the spark plugs of our car refuse to serve if wetted.

In my little booklet on bioenergetics emphasis was laid on the excited states, singlet as well as triplet. This I did because I thought that till we really know what's what we had better look at everything. As men-tioned in my present Chapter IV, a direct transition from the ground state into the excited state within one and the same molecule seems rather unlikely for proc-esses which are not induced by photons. So, in a way, I lost faith in excited states and levels. Charge transfer helped me to recover it.

IONS

The donor-acceptor relations may also lead to a better understanding of ionic activity. Hofmeister has arranged the various ions according to their colloidal activity into the so-called "lyotropic series." This col-loidal activity has been connected since with various parameters, as, ionic radius, the intensity of the sur-rounding electrostatic field, and hydration. For anions the series is mostly quoted as: citrate > tartrate > SO_4

[†] Ref. 6, page 6.
[‡] Woods Hole Conference on Molecular Excitation, 1959.

> acetate > CIO_3 > NO_3 > Br > I > CNS. For cations: Li > Na > K > RB > Cs. The colloidal effects are opposite on the two ends, and accordingly the initial members of the anion series are called "salting out" and the last members the "salting in" anions. My own laboratory has often made use of these activities, "salting out," precipitating proteins by sulfates, or "salting in," bringing into solution and depolymerizing actin, or dissociating actomyosin by elevated concentration of iodides. Rhodanides had similar effects. One may wonder why these ions "salt in," dissolve, dissociate, and whether some other parameter has not been overlooked? Such a parameter may be the ability to act as electron donors or acceptors. This assumption is supported by the ionization potentials and the parallel electron donor properties of the anions. SCN^- and I^- are at the end of the series and are relatively good electron donors. Br^- and $(NO_3)^-$ are less active but even $(CIO_3)^-$ is known as a fair donor.[*] The cations can be expected to act as electron acceptors, though they are less dominating and there is less difference among them, as expressed by their oxidation potentials (Table II). According to the Table, lithium will be the best acceptor, Na will be poorer than K, but there will be very little difference between them. All the same, when studying the action of a salt, we must bear in mind that what we measure in most cases will not be the donor action of the anion nor the acceptor action of the cation, but both, acting simultaneously. So, the solubilizing action of KI on a protein might be due to the prevailing donor activity of the I^-, the solubilizing indicating only the overall result of the dona-

[*] Ref. 47 on page 55.

tion and acceptance of electrons, and will not mean that the cation did not act at all.

TABLE II
OXIDATION POTENTIALS OF ELEMENTS

Element	Oxidation Potential[a]	Element	Oxidation Potential[a]
Li	+3.045	F_2	−2.85
Na	+2.714	Cl_2	−1.36
K	+2.925	Br_2	−1.066
Rb	+2.925	I_2	−0.536
Cs	+2.923		

[a] Values, in volts, referred to the hydrogen-hydrogen ion couple as zero, for unit activities at 25°C. For cations, as Li the reaction is: $Li \rightarrow Li^+ + 1e$, for anions like F the reaction is $2F^- \rightarrow F_2 + 2e$. (Quoted from the Handbook of Chemistry and Physics, 38 Ed., Chemical Rubber Publ. Co.)

The donor activity of iodide can easily be demonstrated by freezing a 10^{-3} M FMN in presence of 10^{-2} or 10^{-3} M KI. On freezing,* the solution assumes a brownish color, indicating the charge transfer with the formation of FMN⁻. If the mixture is frozen in the presence of 1% HCl, the final color reached within a few hours is black. If time is given and the mixture is stored at the low temperature over night, even the isomolar 10^{-3} M KI will make the FMN turn completely black. Such a slow reaction is typical for many charge transfer reactions. The phosphorescence of the frozen FMN is completely wiped out, indicating the transfer of a whole electron, that is a strong charge transfer reaction in the ground state. KSCN also turns the FMN solution brownish, but seems to be less active

* See first footnote on page 77.

than KI, which difference may be due to the poorer approachability of SCN^- and the greater "softness" of the I^- ion. That KSCN actually is a stronger donor than KI can be shown by using a different acceptor, and freezing a 10^{-3} solution of 1, 2-naphthoquinone-4-sulfate Na in presence of 0.1–0.01 M KI or KSCN. The naphthoquinone radical is brown, and on freezing the solutions turn brown. The tube containing KSCN will be darker brown than the one with KI, all these reactions being reverted on melting.

Charge transfer complexes of FMN are split by 0.1 M "neutral" salts. This holds true for many other charge transfer complexes. This dissociating action was, for a long time, a mystery to me, for heteropolar properties should rather promote than inhibit charge transfer. However, if anions and cations act as electron donors and acceptors, then this dissociating action can be ascribed to a competition of the anion and cation of the salt for the acceptor, respectively the donor of the complex.

METACHROMASIA

Metachromasia is a fascinating phenomenon, in all probability also connected with charge transfer. By metachromasia, following Ehrlich,[108] is meant the ability of certain dyes to stain different material in different color. Metachromasia has been studied by several investigators, most intensely by Michaelis and Granick.[109] As they showed, very dilute solutions of metachromatic dyes (over 10^{-6} M) have a sharp ab-

[108] P. Ehrlich, *Arch. mikroskop. Anat. u. Entwicklungsmech.* **13**, 1877.
[109] L. Michaelis and S. Granick, *J. Am. Chem. Soc.* **67**, 1212, 1945.

98

sorption peak, given by the monomer. If the concentration of the dye is increased another lower peak appears at a somewhat shorter wavelength, which is generally accepted to be that of the dimer of the dye[110,111] (Fig. 17). With increasing concentration the

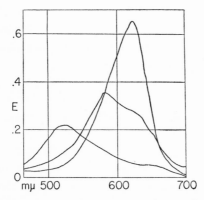

FIG. 17. Highest curve: Absorption of 10^{-6} M toluidine blue, 10 mm path. Lower curve: 10^{-5} M toluidine blue, path 1 mm. Lowest curve: same plus equivalent polyethylenesulfonate Na.

monomer peak becomes lower, the dimer peak higher. The latter always remains considerably lower than the monomer, even if dimerization is nearly complete. If a substance is added which is stained metachromatically, a third, still lower and broader structureless peak appears at a shorter wavelength and both the monomer and dimer peaks disappear (Fig. 17). The classic metachromatic dyes are toluidine blue, methylene blue, and thionine.

[110] There is also a smaller peak and shorter wavelength. Th. Förster, *Naturwiss.* **33**, 166, 1946.
[111] J. Lavorell, *J. Phys. Chem.* **61**, 1600, 1957.

Lison[112] observed that most naturally occurring substances which color metachromatically such as agar-agar or chondroitin, contain SO_3 groups and if these groups are eliminated the metachromasia disappears. Evidently, the metachromatic coloration had to be due to an interaction of the SO_3 group with the dye. Since the absorption shifts to shorter wavelength and becomes smaller on dimerization, it was natural to believe that metachromasia, a further shift and decrease, was due to a further association of the dye, induced, somehow, by the SO_3 groups of the stained substance. The theory of polymerization (suggested with a question mark by Michaelis) was lately revived by Bradley and Wolf.[113]

The naturally occurring metachromatically dyeing substances, such as agar-agar or chondroitin, owing to their complex and only partially known structure, are unfit for quantitative study. The same holds for polyphosphate, polymetaphosphates, or silicates, which also give metachromasia. Thanks to the courtesy of the Upjohn Company, Kalamazoo, I am in possession of a polyethylsulfate, which is an almost ideal material. It consists of repeat units of CH_2—$CHSO_3Na$, has a molecular weight of 12,900 and is soluble in water.

If one of the metachromatic dyes, say toluidine blue, is gradually added to a stronger solution of the polyethylsulfate ($10^{-2} M$ in relation to the ethyl sulfate unit which weighs 134 grams) the first minute quantity of the dye which produces a visible color or meas-

[112] L. Lison, Histochimie Animale, Gauther, Paris, 1936; *Arch Biol.* **46**, 599, 1935.
[113] D. E. Bradley and M. Wolff, *Proc. Natl. Acad. Sci. U.S.* **45**, 944, 1959.

urable spectrum is metachromatic. This simple observation strongly speaks against the polymerization theory of metachromasia because the few molecules of the dye added can be expected to disperse rather than aggregate, there being less than one dye molecule per macromolecule.

The energy levels of the three classic metachromatic dyes mentioned show a peculiarity which I did not find in the several scores of dyes studied in collaboration with G. Karreman. The k's, both those of the highest filled and those of the lowest empty orbital, are very small (Table III) which makes the gap between

TABLE III

(condensed from Table I)

k VALUES FOR THE HIGHEST FILLED (hf) AND LOWEST EMPTY (le)
MOLECULAR ORBITALS

	hfmo	lemo
Toluidine blue	0.391	−0.359
Methylene blue	0.398	−0.354
Thionine	0.398	−0.354
Nile blue sulfate	0.591	−0.133

the two very narrow—hence the blue color with its long wave absorption which lies partly in the infrared. These values make the dyes into both good donors and good acceptors. It follows that they should actually be able to form a self-charge transfer complex, one molecule of the dye donating an electron to another, if we bring two molecules into close proximity, as can be done by freezing their dilute aqueous solution. On freezing of their solution all three dyes actually change their color from the "normal" blue to the metachro-

101

matic purple. All this, taken together supports the idea that the metachromatic coloration is due to the formation of a charge transfer complex.[*] SO_3 groups, according to the tabulation of Mulliken,[†] are "ketoid" acceptors while the O's with their lone pairs of electrons could also act as donors.

If one molecule of toluidine blue is added pro SO_3 a metachromatic complex is formed which precipitates. An excess of the colloid, in the relation 2:1, keeps the complex in solution. If more than one molecule of dye is added pro SO_3, a 1:1 metachromatic complex precipitates leaving the excess of the dye in the dark blue solution behind.

There are three points which make metachromasia rather fascinating. Lison[‡] observed that the SO_3 group gives metachromasia only if linked to a macromolecule. Low-molecular sulfo compounds give no metachromasia. I can fully corroborate this statement. My three polyethylsulfate preparations of molecular weight 6000, 12,900, and 27,000 grams reacted equally strongly. However, when going under 5000 grams molecular weight, taking other big molecules such as heparin or chondroitinsulfuric acid, the metachromasia became poorer. The limit seems to lie around 5,000 grams molecular weight. The size of the molecule seems to

---

[*] This does not entail that the metachromatically shifted absorption of the dye corresponds to a charge transfer spectrum. It is, probably, analogous to the shift in the spectrum of iodine in charge transfer reactions. In charge transfer complexes formed with alkaloids or pyridine the spectrum of iodine shifts isosbetically from 520 mμ to 410 mμ while the charge transfer spectrum proper is in the UV.

[†] Ref. 47 on page 55.

[‡] Ref. 112 on page 100.

mean something for charge transfer and it may mean something most important for biology, but what, I am unable to say. The 5,000 gram limit is fascinating because, as judged by the experience on myosin, this is the smallest size unit of which this protein is built.

This relation of metachromasia to molecular weight may have something in common with the experience of Kainer, Bijl, and Rose-Innes, who found that the bulkiness of the reactant molecules favors charge transfer.[*]

The other point which fascinates me is connected with Strugger's[114] observation, according to which live tissues are colored green by acridine orange, and dead tissues, red. According to my experience the situation is not this simple. All the same, metachromatic coloration may open a way to study donor and acceptor properties of live tissues, and the living state or death may have intimate relations to these properties.

The cartilage and synovial membrane is rich in SO_3 groups and one may wonder what their function should be. Two antimalarial agents, atabrine and chloroquine, both of which are good electron donors, have been found to alleviate symptoms of arthritis, alleviated also by cortisone, another donor. These relations might suggest new ideas about this disease which is crippling a considerable percentage of our kind.

The third point which made metachromasia interesting for me is that it may shed light also on the nature of ionic activity. The experiments of B. Kaminer (unpublished) show that salts, containing the two last anions of the "salting in" end of the Hofmeister series,

[*] Ref. 57 on page 60.

[114] S. Strugger, *Jenaische Z. Naturwiss.* **37**, 1941.

induce a metachromatic coloration in all three dyes discussed. This pleads for the correctness of both assumptions: that the salting in effect is due to electron donation and that metachromasia is due to charge transfer. The comparison of the dyes discussed supplies further evidence. As the k values indicate (Table III) Nile blue sulfate is a better acceptor than the other three dyes. Accordingly, we can expect that it is also more sensitive to the action of anions, an expectation borne out by the experiment.

The fact that Nile blue sulfate shows a metachromatic change with nitrate, bromide, iodide, and rhodanide is rendered especially fascinating by the observation of Kahn and Sandow,[115] according to which these four anions are capable of increasing the maximal twitch tension of muscle by their action on the membrane. It seems likely that this action is due to the electron donating properties of these anions, which may open the way to the closer correlation of function, charge transfer, and the understanding of ionic activity and drug action.

An apparent inactivity of a salt, like KCI, does not necessarily mean that its ions did not act as donors or acceptors. It may also mean that these just compensate each other's overall action. So even the ions of an apparently inactive salt may render empty or saturated energy bands conductant by donating or accepting electrons.

[115] A. J. Kahn and A. Sandow, *Ann. New York Acad. Sci.* **62**, 137, 1955.

Part Two

Problems and Approaches

IX

On the Mechanism of Drug Action

IT IS NOT THE INTENTION OF THE AUTHOR TO attempt to give an answer to the unsolved problems outlined at the outset. All he hopes to do in this part of his booklet is to show that some of these problems, if looked at through the glasses of the submolecular, may appear in fresh light, suggesting tentative theories which may lead to useful experimentation.

The first example I want to take is that of the mechanism of drug action. A number of drug actions have found an explanation in competitive inhibition. As a false key may fit into a slot without being able to open it, so a drug may fit into some biological binding site, and displace some natural substance without being capable of fulfilling its function. A few biologically active substances were found to act by blocking some important atomic group, as the SH. In spite of these successes, the action mechanism of the great majority of drugs remained unexplained. This pertains for instance, to most of the alkaloids, and pertains also to the hormones, drugs produced by our own body free of charge.

Drug action represents a wide field of inquiry and

our approach can be but a very limited one. We may ask, for instance, whether charge transfer is not involved in drug action, some of the drugs acting as electron donors or acceptors. It is difficult to predict the biological effect of such a charge transfer, for the action will not only depend on the question, whether electrons are donated or tapped off, but will also depend on the site of action. The cellular membrane, for instance, which dominates many functions of the cell, has mostly a negative charge inside and a positive one outside. So electrons donated at the inside should increase the charge and lead to hyperpolarization, and with it to inhibition, while electrons donated at the outside will decrease the potential and can be expected to cause excitation. The opposite will hold for electron acceptors. We can be prepared also to meet paradoxical results. Let us suppose that a biological substance acts by donating electrons. A drug, acting as electron donor, may compete with a natural substance for its acceptor. Interfering thus with the normal course of electron transmission this drug, though itself a donor, may produce an effect which corresponds to the inhibition of electron donation. In spite of all these incertitudes we may expect one definite interrelation: If a substance exerts its biological activity by accepting or donating electrons, then it should have exceptional acceptor or donor properties.

An indication that acceptor-donor properties may underlie pharmacological action was given on the preceding pages. It was shown that the indoles are exceptionally good donors, and a great number of biologically active substances contain an indole ring (serotonin, lysergic acid, bufontine, indole acetic acid). Attempts were made earlier by Popov, Castellani-Bisi, and M. Craft[116] to connect pharmacological activity, like convulsant effect, with electron donation. That indoles may actually act in their biological reactions as electron donors is further suggested by the fact that the OH group, induced into the serotonin molecule at position 5 equally increases pharmacological activity and the electron donating property. Serotonin is one of the strongest donors I ever met though the k of the highest filled orbital is but moderately lower than that of indole (0.461). So there are still unidentified molecular parameters which greatly influence donor-acceptor properties. That charge transfer may be involved in pharmacological activity is suggested also by the fact that the various alkaloids, such as quinine, nicotine, atropine, physostigmine, morphine, strychnine, aconitine, etc., all act as good donors toward iodine in an anhydron solution,* and so do the ketosteroids, while estrogens, as mentioned before, prove to be donors toward trinitrobenzene. An additional evidence for the possible pharmacological importance of acceptor-donor properties was given by Fujimori† who showed that various pteridine derivatives are the stronger antimetabolites of folic acid the stronger they act as acceptors toward tryptophan.

---

* The I_2-spectrum absorption (in chloroform solution) shifts in presence of these alkaloids with an isosbestic point from 520 mμ to a shorter wavelength around 400 mμ, while in the UV a broad charge transfer is developed. With nicotine the charge transfer spectrum develops slowly, within an hour or so. This tardiness of the reaction is characteristic for many charge transfer reactions.

† Ref. 94 on page 86.

[116] A. I. Popov, C. Castellani-Bisi, and M. Craft, *J. Am. Chem. Soc.* **80**, 6513, 1958.

As a further example we may quote nitrophenols, in the first place 2,4-dinitrophenol and halophenols. Dinitrophenol is known to have a strong uncoupling activity on oxidative phosphorylation, one of the greatest puzzles of current biochemistry. Polynitrophenols, on the whole, are very good electron acceptors, owing to the unsatisfied electronegativity of their nitro group, and it is believable that they capture the high-energy electrons which should go into the "black box" discussed on page 23, forming charge transfer complexes with their donor. The same holds for halophenols, the strong activity of which, on phosphorylation, has been discovered by Clowes and Krahl.[117-119] Their action is less specific than that of dinitrophenol (Middlebrook et al.[120]). It should also be remembered that one of the most important regulators of cellular oxidation, the hormone thyroxine, is also a halophenol, containing the most electronegative halogen, iodine. It uncouples oxidative phosphorylation. It develops its activity very slowly, in 48 hours or so—a striking sluggishness, characteristic of many charge transfer reactions, one example of which has been analyzed by R. H. Steele, i.e. the very slow complex formation between rhodamine B and dichlorophenol.* Nitro compounds also promote the transition of the attracted electron into

* See "Bioenergetics," (A. Szent-Györgyi, Academic Press, New York, 1957), page 54, notes about the depressant action of dichlorophenoxyacetic acid on oxygen uptake; see Ibid. p. 122.)

[117] G. H. A. Clowes and M. E. Krahl, J. Gen. Physiol. 20, 145, 1936.
[118] M. E. Krahl and G. H. A. Clowes, Ibid. 20, 173, 1930.
[119] M. E. Krahl and G. H. A. Clowes, J. Cellular Comp. Physiol. 11, 1, 21, 1938.
[120] M. Middlebrook and A. Szent-Györgyi, Biochim. et Biophys. Acta. 18, 407, 1955.

the triplet state.† This can be expected also from thyroxine with its very heavy iodine atoms.

Another train of thought was followed by B. Kaminer (unpublished). The proboscis (a smooth muscle preparation of *Phascolosoma* (a marine worm), if denervated, shows no spontaneous rhythmicity. Serotonin, a donor, is inactive. However, if the preparation is treated first with good acceptors, serotonin induces rhythmic contractions. The reverse, also, holds: after treatment with serotonin, acceptors induce rhythmic contraction.

The obvious way to find additional evidence for a connection between electron transfer and pharmacological activity could be found by picking drugs with extraordinary pharmacological activity and then asking whether they have extraordinary qualities as electron donors or acceptors. If the extraordinary biological activity of the drug in question is due to an electron transfer then we can also expect that drug to have exceptional qualities as electron donor or acceptor.

A drug with a unique biological activity is the tranquilizer chlorpromazine* (Fig. 18) which is widely used in treating the symptoms of schizophrenia. Since its introduction a great number of hospital beds have become empty, because among all diseases it was schizophrenia which had permanently occupied the most hospital space. With this end in view, Karreman *et al.*[121] calculated the *k*-values of this substance and

† Ref. 52 on page 59.
* The author is grateful for a supply of this substance "Thorazine" to Smith, Kline and French, Philadelphia, Pa.

[121] G. Karreman, I. Isenberg, and A. Szent-Györgyi, *Science* **130**, 1191, 1959.

111

found that of the highest filled orbital quite exceptionally high. In fact, it is smaller than zero, has a negative sign, being equal to −0.210, the orbital having an antibonding character. Such values had been found earlier by B. and A. Pullman in leucomethylene

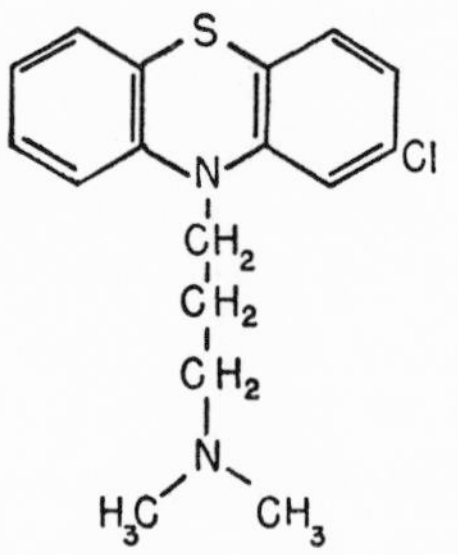

FIG. 18. Chlorpromazine.

blue or reduced riboflavine, but both these substances are unstable, autooxidize rapidly, two electrons having been forced upon them. Chlorpromazine is the first substance found which has an antibonding highest filled orbital in its normal, stable state. It can be expected to be an exceedingly good monovalent electron donor capable of forming stable charge transfer complexes, the complex formation being supported by the planar nature of the extensive system of conjugated double bonds with its pool of π electrons and the lone pairs of electrons on the N and S atoms.

These ideas may become the starting point of various trains of thought. If the action of this drug is due to its qualities as electron donor and these qualities can be calculated, then, maybe, new and even more potent drugs may be found by calculation, which

112

may be still better electron donors and may have an even more favorable action than chlorpromazine. And, if the symptoms of schizophrenia can be influenced favorably by electron donation, then, perhaps, a lack of electrons may be involved in the genesis of this disease, and if so, the question comes up: what has induced it? If the disturbance can be corrected to some extent by electron donation, could it not have been caused by the presence of an electron acceptor? It is most fascinating to note, in this connection, that B. Pullman and A. M. Perault[*] find hematoporphyrins both good donors and good acceptors, and find the metabolic product of protoporphyrins, biliverdin, a quite exceptionally good electron acceptor. This substance is, in a way, the counterpart of chlorpromazine, the k of its lowest empty orbital having a plus sign (0.021), a bonding character. To my knowledge this is the first substance found to have such a value. Urobilin, a further metabolite has still strong donating properties. What makes these data exciting is the fact that we metabolize a great quantity of hem daily, 0.7% of our total blood, and Klüver[122-125] has for many years called attention to a close relation of hematoporphyrin and mental disturbance. In newborns a hyperbilirubinemia is often found to be accompanied by grave anatomical damage to the brain. Bilirubin inhibits the

[*] Ref. 4 on page 6.

[122] H. Klüver, Science 99, 482, 1944.
[123] H. Klüver, J. Psychology 17, 209, 1944.
[124] H. Klüver, Functional difference between the occipital and temporal lobes, in "Cerebral Mechanisms and Behavior," L. A. Jeffres (Ed.), Wiley and Son, New York, 1951, pp. 172–178.
[125] H. Klüver and E. Barrera, J. Psychol, 37, 199, 1954.

production of hems,[126] and similarly to several other electron acceptors, inhibits oxidative phosphorylation.[127] Should schizophrenia be due to the presence of a strong acceptor in the blood, then, maybe, this substance could be inactivated by a strong donor with which the acceptor could form a charge transfer complex, relieving the patient of all his troubles. I am not proposing here a new theory of schizophrenia. What I am trying to show is merely that quantum mechanics may suggest unexpected new approaches to important problems which have been stagnant such a long time. The examples quoted also show that the distance between those abstruse quantum mechanical calculations and the patient's bed may not be as great as believed. Starting on these lines, schizophrenia, with all its mysterious psychic aberrations, may turn out, some day, to be but the side effect of some, in itself harmless, metabolic disturbance which, once recognized, can perhaps easily be corrected.

[126] R. F. Labbe, R. Zaske, and R. A. Aldrich, *Science* **129**, 1741, 1959.
[127] L. Ernster, L. Herlin, and L. Zellerstrom, *Pediatrics* **20**, 647, 1947.

X

On ATP

ATP IS THE MAIN FUEL OF LIFE PRODUCED IN photosynthesis and oxidative phosphorylation. In both cases it is produced by an electric current, that is, the energy released by a "dropping" electron. So the question arises whether the energy of the $\sim$ is not transformed into a current back again when it has to drive life and produce work, w. I have confessed earlier my inability to make sense of the transformation of chemical energy into w, except by supposing that, somehow, the chemical potential is translated into electronic energy.

A transformation demands a transformer. Looking at the structure of the ATP molecule (Fig. 19) one wonders whether the adenine group is not the transformer. Why should Nature have attached it to the phosphate chain if she only wanted the energy of $\sim$'s? Pyrophosphate or inorganic triphosphate should have done just as well! I have also shown, in my little "Bioenergetics,"* that the phosphate chain and the adenine are linked together by the pentose in such a way that

* "Bioenergetics," page 64.

the former can fold back on the latter making the two terminal phosphates touch upon the two N's at positions 6 and 7 and form a tetradentate chelate with them, using Mg as a link. This possibility has found some, though not conclusive, evidence in the infrared studies of Epp, Ramasarma, and Wetter.[128] The phosphates may fold back onto the adenine also in such a way as to lie flat on the face of the purine. In this case the two terminal phosphates could be in touch with the ring but not the third one. The folding could take place in the pentose and so need not bend the phosphate chain, which bending would be resisted by the negative charges on the chain and would also interfere with the intimate contact with the purine. Adenine is a good donor and the P atom has unoccupied d orbitals which could make it into an acceptor.* So the possibility exists that within the ATP molecule an electron is transmitted from adenine to phosphate, either from the former's π-pool, or from the lone pair of one of its N's.† The rotatory dispersion of ATP solutions also indicates a folded molecule.

* It seems possible to give reasons why Nature may have selected the P atom for its central role. Elements on the left side of the periodic table are metallic and tend to give off electrons and not take them up. The middle ones, C and Si have no special affinity, while the ones farther to the right of P may have a too great electron affinity to serve as reversible electron transmitters. Of the elements in the same column as P, N would not do since it has no d orbitals. The next heavier element, As, seems to be incompatible with life for other reasons. So P may be the only element which could fill the role of a reversible electron transmitter. I am indebted to Dr. M. Calvin for discussions and ideas on this line.

† The metachromasia of polyphosphates supports the assumption of their being good electron acceptors.

[128] A. Epp, T. Ramasarma and L. R. Wetter, *J. Am. Chem. Soc.* **80,** 724, 1958.

It seems worthwhile to reflect about this point since this formation, adenine-pentose-phosphate, returns in almost all important biological catalysts of energy production or transmission: DPN, TPN, FAD, and coenzyme A. The first three have an analogous formation also on the other side of the phosphate chain, the sequence being pyridine-pentose-pyrophosphate-pentose-adenine, or isoalloxazine-pentose-pyrophosphate-pentose-adenine, so that not only the adenine could fold back onto the phosphates but also the pyridine and isoalloxazine. The phosphate thus sandwiched between adenine and pyridine, respectively adenine and iso-

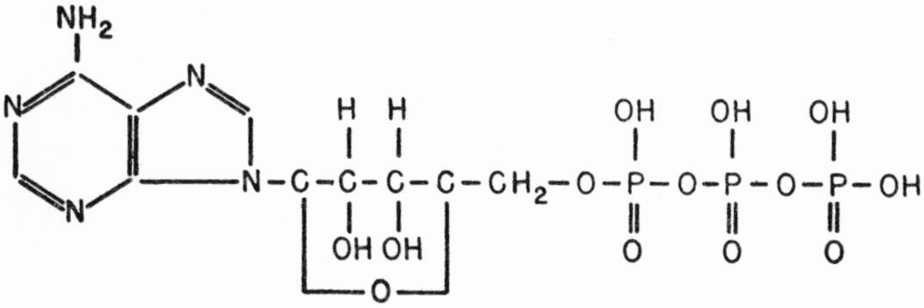

FIG. 19. Adenosinetriphosphate (ATP).

alloxazine could mediate the electron transfer between the two cyclic structures on its two sides. Weber actually showed a strong interaction between adenine and isoalloxazine[129] respectively pyridine.[130] In coenzyme A there is no pentose between the phosphates and pantothenic acid, but this latter has enough freely rotating links to be able to fold back on the phosphate. It may be also worth mentioning that only triphosphate or diphosphate can fold back onto the adenine

[129] G. Weber, *Biochem. J.* **47,** 114, 1950.
[130] G. Weber, *Nature* **180,** 1409, 1957.

117

in a way to allow a sufficient overlap and charge transfer. Monophosphate would not do, as is easy to show by means of an atomic model. All the catalysts mentioned contain triphosphate or diphosphate (pyrophosphate), but not monophosphate.

A charge transfer involves the uncoupling of an electron pair, and represents thus, in a way, an embryo of a current, generating a negatively and a positively charged molecule. So we may turn our phantasy loose for an instant, and ask whether such a charge transfer could help us to explain the function of these nucleotides? Let us consider first the synthesis of ATP from ADP and P in photosynthesis. The Pullmans* have shown lately that, almost without exception, the two atoms disconnected in enzymic hydrolysis carry a positive formal charge. This is true also for the P—O bonds in ATP. These charges can be expected to predispose the "dipositive bond" for hydrolytic splitting by their mutual coulombic repulsion. But, if these charges predispose the P—O bond for splitting, they must also resist the formation of this bond, and so an electron, transferred onto the P by charge transfer must turn the repulsion into an attraction promoting synthesis. The empty place on adenine may be filled by an electron of the oxidative cycle, releasing thus the electron donated to the phosphate chain from its electrostatic bondage.

For consideration of the reverse process, the splitting of the P—O—P bond and the transformation of its energy into a current, we could take as an example muscle contraction in which the energy of the ~ is transformed into motion—mechanical work. Let us

* Ref. 5 on page 6.

118

start with an ATP molecule in which an electron has been transferred from adenine to the phosphate chain.* By this transfer the adenine has to become a good acceptor, the phosphate a good donor, and so (as a *jeux d'esprit,* a play of the mind) we could suppose that adenine accepts an electron from myosin while phosphate donates one to actin. This would make the positive and negative charges become widely separated. Now let us imagine that we take the myosin molecule between the two fingers of our right hand, the other end (actin) between the two fingers of our left, and pull the chain apart. If we do this slowly the transferred electron can be expected to slip back to its original position. If we would tear the chain in two very fast, allowing no time for the electrons to slip back, we would have to do work to overcome the coulombic attraction between positive and negative charges. We could tear up the complex without resistance only in one case: if, previously, we have hidden in the link to be torn a sufficient amount of free energy to pay for the process. In this case the chain could be broken without external energy. The P—O—P bond actually has about 10 kcal worth of energy hidden in it and so this bond could simply break and thus separate the positive and negative charges in actin and myosin, generating thus an electric potential, a current. There is no difficulty in making theories of how such a current could produce contraction.

I can see here the outlines of an extremely cunning

---

* Owing to the unshared pairs of electrons of the O atoms the phosphate chain can be looked upon as conjugated with mobile π electrons.

119

scheme. Possibly, Nature created in ATP a substance which could take up an electron on one end and give one off on the other, while the terminal $\sim$ has enough energy hidden to make it possible for a hydrolytic enzyme to split the bond, thus trapping the positive and negative charge out in the contractile protein, thus transforming chemical energy into a current.[*]

I am conscious of speculation which, at the moment, has no experimental support. All the same, it is this speculation which allows me to make for the first time, after twenty years of muscle research, an intelligible picture of how ATP could drive muscle. This is a most fundamental problem, close to the core of life.

It can be hoped that ESR will give the answer to these problems. It is impossible to predict whether a charge transfer between adenine and phosphate would give a signal or not, depending on the degree at which the two electrons become uncoupled. All the same, there is no reason for not putting ATP into the ESR machine. This has been done by Isenberg et al.[134] with

---

[*] It seems possible that some such tricks, as represented by the hypothetical function of ATP in muscle, has been applied by Nature in other fields too. So, in vision, the photon absorbed by retinene may cause a charge transfer onto opsin. The *cis-trans* transformation which follows the photochemical acts, splits the link between dye and protein, or at least may interfere with the overlap of wave functions, trapping, thus, the transferred electron on the protein, starting up "excitation."[131–133]

131 G. Wald, *Scientific American,* **201,** 92, 1959.
132 V. J. Wolff, R. S. Adams, H. Linschitz, and D. Kennedy, *A.M.A. Archives of Ophthalmology* **60,** 695, 1958.
133 V. J. Wolff, R. S. Adams, H. Linschitz, and E. W. Abramson, *Ann. New York Acad. Sci.* **74,** 281, 1958.
134 I. Isenberg and A. Szent-Györgyi, *Proc. Natl. Acad. Sci. U.S.* **45,** 1232, 1959.

the help of Dr. H. Kallmann of New York University, who kindly offered his equipment. A signal was obtained, consisting of a broad peak, stretching over several hundred gauss and carrying smaller secondary peaks. Since that time the author's laboratory has become the happy owner of an ESR machine. On this machine ATP is now subjected to a more detailed study, allowing us to experience all the difficulties connected with ESR studies. We hope to learn soon to what extent the observed spectra were artifacts, or due to impurities, or were due to the electronic distribution in the ATP molecule itself, giving information about the nature and function of this molecule.

XI

On the Chemistry of the Thymus Gland

RESEARCH IS RARELY GUIDED BY LOGIC. IT IS GUIDED mostly by hunches, guesses, and intuition. All the same, once we get somewhere and present our results, we like to present them as a logical sequence. If the course of our work had to be plotted, we would plot it as the straight line in Fig. 20, where A follows B, B

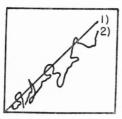

FIG. 20. Curve 1: course of research as presented. Curve 2: actual course. (See text.)

follows C, etc. The real curve would be more like curve 2, at least for my studies on the thymus.

In my book on "Bioenergetics" I dragged in the thymus without any binding reason and tried to show

123

how logically it fitted into my story, though it was but a hunch that this gland had something to do with energy transmission because its extracts were colored more yellow than corresponded to their FMN content. I only guessed that myotonia was due to a defective energization of the membrane and that the absence of the "thymus substance" was responsible for it. However loose my reasoning, I could show that the thymus extracts actually cured the myotonic symptoms in goats. As is generally known, 2,4-dichlorophenoxy-acetic acid produces myotonic symptoms in animals which might be due to the drugs complexing with my hypothetical "thymus substance." So I was encouraged to find that the active agent which benefited my goats, could be shaken out from the watery thymus extracts by 2,4-dichlorophenol.

I am still unable to state what this active substance is, or tell whether it is a specific product of the thymus. If I bring this gland in again, it is for two reasons. In my "Bioenergetics" I left the story unfinished. Although I am still unable to finish it, I feel the obligation toward my earlier readers to tell them that I am still trying to finish it. My second reason is that my observations brought to light the existence of a second substance which is antagonistic to the first, and it is not impossible that the two represent some basic biological balance connected with the acceptor-donor relation. So, even if it does not clear up the physiology of the thymus, my work can be quoted at least as an example of "serendipity," the experience of looking for one thing and finding another.

To be correct, this work on the thymus is not my work at all, but a joint work, started in association with Jane McLaughlin, and pursued at present in association with Dr. Andrew Hegyeli, while the animal tests are conducted by Dr. Hedda Rév.

The difficulty in isolating the active substance lay in the fact that we possessed no suitable biological test for it.* Looking around for a bioassay we also tried its influence on malignant growth. Our extracts seemed to slow it down, or stop it altogether.

I will forego the discussion of the difficulties of using malignant growth as a test object. Eventually the difficulties were solved by using one and the same animal for the test and its control, measuring growth rate while under the influence of extracts, and after. Deviation from logarithmic growth was found to be a reliable indication of an activity.

With this test in hand, isolation was begun. For a little while things went well. Then everything got mixed up and activity decreased or was lost where it had to increase. Eventually the puzzle was solved by the demonstration that our extracts contained two active substances, the one promoting, the other inhibiting malignant growth and the result depended on their balance. It is a common experience that two unknown variables make results messy.

The problem is still vigorously pursued and it is hoped that the next edition of this booklet (if any) will contain the chemical formulas of the two active agents.

* Myotonic goats, as a test object, are very unsatisfactory.

XII

The Living State

THIS, TO MY MIND, IS A MOST INTRIGUING PROB-
lem. It is also the most obscure one, as difficult to define as life itself. Although we cannot define life, we
know life from death and can distinguish between a
dead cat and a live one, which corresponds to the two
basic states of biological systems. The problem is, perhaps, not quite as abstruse as it appears on first sight
because we can produce the transition by simple experimental means from the one state into the other, at
least "one way." To stay with the cat, we can, for instance, wipe out consciousness in one blow by clamping
the carotid artery, that is, cutting the O_2 supply to
the brain. Consciousness being the main product of the
brain, this means the cessation of biological activity.
It could be objected that this is not death, the change
being reversible. However, the reversibility has very
narrow limits. If we keep the carotids clamped for a
few minutes, the change becomes irreversible. This
means that the living system was in a metastable state
which needed a permanent supply of energy for its
maintenance, the half-lifetime of the living state being

127

of the order of minutes. No doubt, similar changes can be produced also in other organs though the half-lifetime may vary to some extent. The same point can be demonstrated by cyanide in an even more dramatic fashion. If we inject cyanide into one of our veins the first thing we notice is that we are dead. Cyanide, as we know from Warburg's classic work, cuts out the activation of O_2, combining with some catalytic metal present.

That living systems are in a metastable state which demands the permanent supply of energy for its maintenance, is no surprise. The second law of thermodynamics could predict this. What is unexpected is the brevity of the half-lifetime. This is a surprise because most organs have a considerable store of ATP, backed up by creatine phosphate, while limited amounts of ATP can also be produced anaerobically. This ATP should be able to tide us over much longer periods of lack of O_2, if it could supply the energy needed to maintain the metastable living state.

There is no doubt in my mind that the energy for muscular contraction is derived from ATP. When muscle enters into activity to perform its public function, contraction, it suddenly demands great amounts of energy. O_2 supply is slow and continuous. So muscle could not depend on O_2 for the energy of its contraction, directly, and has to use as energy source a substance like ATP, available any time in moderate amounts. But, if ATP can cover the sudden great energy demands of a public function, why can it not cover the modest and continuous demands made by the private life of the cell, the maintenance of its metastable state? Is it not that there are two inde-

128

pendent systems of energy production, both using O_2 as their final electron acceptor, the one located in mitochondria and responsible for the production of ATP, while the other is located in the basic cellular structures themselves which have to be maintained in their peculiar state?

A glance at Fig. 4 makes such an assumption seem reasonable. This figure intends to show that a considerable portion of the electrons, which are sent down the chain of oxidative phosphorylation to produce ATP by their energy, are derived from DPNH or TPNH. But why should the organism take this long detour over mitochondrial oxidative phosphorylation and ATP to satisfy a slow and continuous demand? Why could the high-energy electrons of DPNH or TPNH not be placed more directly on the living structure which eventually could couple them to O_2 using their energy more directly?

All this may be speculation, but once we recognize the existence of a metastable living state we have to ask questions and speculate to find a reasonable working hypothesis. To say that it is ATP which maintains the living state is no more than a speculation either, a speculation burdened with major difficulties. The hypothesis I am proposing is this: the cell derives the energy necessary for the maintenance of its living state directly from DPNH or TPNH. As a first move in this direction, we may ask questions about the two ends of the hypothetical process: how are the electrons of DPNH or TPNH transmitted to the protein edifice, and how could electrons be transmitted from this edifice to O_2? There are tentative answers possible to both questions.

129

As to the first: Talalay, Williams-Ashman, and Hurlock[135-138] have shown that steroids can mediate the electron transfer between pyridine nucleotides in the presence of protein (both specific and unspecific). Why then could they not mediate the electron transfer between TPNH or DPNH and the protein edifice itself? That steroids can, with their =O act as donors, has been shown to be probable in Chapter VII, and a steroid which has donated an electron must be a good acceptor, enabling the molecule to act as an electron transmitter.

As to the other end, O_2, Debye and Edwards[139] have found that proteins, illuminated by shorter UV, emit, at low temperatures, a long-lasting afterglow, which has been studied also in my laboratory[140] using, by preference, proteins of the lens which contain no hem dyes. This light emission is completely quenched by O_2 which, evidently, picks up the excited electrons. The direct transmission of electrons from the protein to O_2 may thus be demonstrated by simple means.*

* It is noteworthy that the long-lived afterglow of tryptophan is not quenched by O_2 while that of protein is, though the emission of protein is, in all probability, an emission of its tryptophans. Though less probable, the possibility should not be disregarded that the quenching by O_2 is due to its paramagnetism, being a "paramagnetic quenching."

[135] P. Talalay and H. G. Williams-Ashman, *Proc. Natl. Acad. Sci. U.S.* **44**, 15, 1958.
[136] P. Talalay, B. Hurlock, and H. G. Williams-Ashman, *Ibid.* **44**, 862, 1958.
[137] B. Hurlock and P. Talalay, *J. Biol. Chem.* **234**, 886, 1958.
[138] B. Hurlock and P. Talalay, *Arch. Biochem. Biophys.* **80**, 468, 1959.
[139] P. Debye and J. O. Edwards, *Science* **116**, 143, 1952.
[140] A. Szent-Györgyi, *Biochim. et. Biophys. Acta.* **16**, 167, 1955.

All the same, the experience with cyanide indicates that some metal-containing catalyst is involved in coupling the electrons to O_2.[141]

The hypothesis presented would also give an explanation for the hitherto unknown mechanism of action of steroids. What we know about it at present is hardly more than that steroids are indispensable for life, and life simply fizzles out in their absence.

What happens between the hypothetical steroid and O_2 end is but a question mark, the whole problem having never been stated before. The fate of the electron within the protein or nucleoprotein structures would depend to a great extent on the nature of these macromolecular aggregates and their hydrate shells, whether they are semiconductors, proton conductors, etc. Our approach to these problems will depend to a great extent, also, on the question of what we mean by energy. Whatever this word may mean, the energy of an electron can conveniently be expressed in terms of ionization potential.* It seems likely to me that as

---

* One could also talk about an "electron pressure" (in analogy to pH) which is inversely proportional to IP. The negative potential of the "inside" of most cells, as compared with the "outside" may have relations to the higher electron pressure inside.

[141] The cyan-sensitive enzymes may be peroxidases. H. G. Williams-Ashman, M. Cassman, and M. Klavins have shown (*Nature* **184**, 427, 1959) peroxidases to oxidize DPNH or TPNH under catalytic influence of estrogens. This is the more remarkable since estrogens, which induce a hypertrophy of the uterus muscle, also cause peroxidases to appear in this organ in high concentration (F. V. Lucas, H. A. Neufeld, J. G. Utterback, A. P. Martin, and E. Stotz, *J. Biol. Chem.* **214**, 775, 1955). These peroxidases could also utilize the H_2O_2 formed in the reduction of O_2. That estrogens, with their aromatic structure can act also as π donors, was demonstrated by Williams-Ashman (see page 90).

pH is equalized within the single cells by buffers, so the ionization potentials also are equalized by electron donors and acceptors. Ascorbic acid, with dehydro-ascorbic acid, may be one of these "IP-buffers." Ions, acting as donors or acceptors may also be instrumental in this equalization. As electron acceptors and donors are, in the extended ideas of G. N. Lewis, acids and bases, so "IP-buffer" means but an extension of the idea of an acid-base buffer. The NH groups of the protein backbone can be expected to be strong donors, while the CO groups may act as "ketoid acceptors" or lone-pair donors, both located strategically in the continuous chain of H-bonds which conjugate the whole protein molecule and may create continuous energy bands (see Evans and Gergely* and Eley *et al.*).†

Certainly, the living state involves many factors and at present we can hardly do more than to try to collect single items in the hope of fitting them together later.

One of the characteristics of the living state is the accumulation of ions against a gradient, concentrations becoming equalized in death. We still have no final answer to the question of how ions are accumulated. We only have theories. Is it a redox pump, as suggested by Conway,[142] or is it some different juggling with carriers, or is it a bulk property,[143] a consequence of structure? We do not know. The fact that ion trans-

* Ref. 31 on page 51.
† Ref. 32 on page 51.

[142] E. J. Conway, *Science* **713**, 270, 1951.
[143] S. L. Baird, Jr., G. Karreman, H. Müller, and A. Szent-Györgyi, *Proc. Natl. Acad. Sci. U.S.* **43**, 705, 1957.

port against a gradient is paralyzed by cyanide but not by 2,4-dinitrophenol (which stops ATP production by oxidative phosphorylation) also suggests that the energy used is derived from some other source than ATP.

One important characteristic property of the living state seems to be its paramagnetic behavior, as suggested by the work of Commoner[*] and his associates who found the intensity of the ESR signal given by various animal or vegetable tissue proportional to the intensity of metabolism. One is reminded of the Latin saying: "the faster a motion the more it is."[†] The intenser life and metabolism, the more life it is, and the intenser the paramagnetic behavior, while in death diamagnetic susceptibility increases.[144] The ESR signals of live tissue seem to come chiefly from the free radicals formed in metabolism. Commoner's groups, as well as Ehrenberg and Ludwig,[‡] found the free radical formed on partial reduction of FMN and its protein complexes to give a signal. Other free radicals may have contributed too, and so might have contributed other charge transfer complexes going into their ionic state. In any case, the ESR signal seems to be a signal of life, though given also by stable substances such as melanin or various resins.[**] Anything alive seems to give a signal, "your finger or the mouse's tail."[§]

[*] Refs. 7, 8, 9, 10 on page 6.
[†] "Quo celerior motus eo magis motus."
[‡] Ref. 89 on page 82.
[**] Ref. 7 on page 6.
[§] Conversation with M. Calvin.

[144] E. Bauer, *Nature* **138,** 801, 1936.

These observations give the impression that the paramagnetic behavior is due to the free radicals formed in the one-electron redox processes, but we should not forget that a not inconsiderable part of the living protein structure must actually be present in the form of a charge transfer complex, is thus, in a way, a free radical. To convince oneself of this, one has to look only at the intense brown coloration of the liver. In spite of the undoubtedly numerous efforts, nobody ever has isolated the dye which could have been made responsible for this color. I have isolated a considerable quantity of a chocolate brown substance, which on treatment with HCl, fell into a colorless protein and a golden yellow flavine. Evidently, this substance was a charge transfer complex of protein with FMN and the brown coloration of the liver has to be ascribed to the formation of a charge transfer complex with isoalloxazins. The kidney and the adrenal cortex are brown too. The brain may be white because of the great bulk of dielectrics it has to contain, although the cortex, which is richer in cells, has a brownish coloration, and where cells aggregate, as in the "red nucleus," the brown color becomes quite evident. It is not impossible that flavines are not the only substances forming charge transfer complexes with the proteins. Steroids, as donors, might do likewise, giving charge transfer complexes with spectra in the infrared or ultraviolet.

As the last item on my list of the peculiarities of the "living state" I would like to touch upon one of the most intriguing observations of the last years, that of the very broad signal given by nucleoproteins (Blumenfeld, Kalmanson, and Shen-Pei*). This signal (if really due to the protein-nucleic acid complex) indicates a density of unpaired electrons which is almost comparable to that found in metals, to which metals owe their conductivity. This finding, if corroborated, may lift a veil which now obscures the real nature and meaning of protein, nucleic acid, and nucleoprotein.

All these factors are parts of the magnificent edifice of life, as bricks, lying on the roadside may have been parts of a Greek temple.

In an earlier chapter I emphasized the biological importance of "organization," by which I meant that if Nature puts two things together a new structure is born which can no more be described in terms of the qualities of its components. The same holds also for functions. In living systems the various functions, too, seem to integrate into higher units. We will really approach the understanding of life when all structures and functions, all levels, from the electronic to the supramolecular, will merge into one single unit. Until then our distinguishing between structure and function, classic chemical reactions and quantum mechanics, or the sub- and supramolecular, only shows the limited nature of our approach and understanding.

* Refs. 71, 72 on page 71.